T0318753

Modern Treatment Strategies for Marine Pollution

Modern Treatment Strategies for Marine Pollution

P. SENTHIL KUMAR
Department of Chemical Engineering,
Sri Sivasubramaniya Nadar College of Engineering,
Kalavakkam, India

Elsevier
Radarweg 29, PO Box 211, 1000 AE Amsterdam, Netherlands
The Boulevard, Langford Lane, Kidlington, Oxford OX5 1GB, United Kingdom
50 Hampshire Street, 5th Floor, Cambridge, MA 02139, United States

British Library Cataloguing-in-Publication Data
A catalogue record for this book is available from the British Library

Library of Congress Cataloging-in-Publication Data
A catalog record for this book is available from the Library of Congress

ISBN: 978-0-12-822279-9

For Information on all Elsevier publications
visit our website at https://www.elsevier.com/books-and-journals

Publisher: Candice Janco
Acquisitions Editor: Louisa Munro
Editorial Project Manager: Alice Grant
Production Project Manager: Bharatwaj Varatharajan
Cover Designer: Victoria Pearson

Typeset by MPS Limited, Chennai, India

Contents

Foreword

The ocean plays a key role in maintaining the ecological balance of the carbon, nitrogen and phosphorous cycles and a variety of important chemicals. It is disturbed when external matter mingles with marine ecology and its environment. This external matter can be pollutants produced by human activities. If these activities are not controlled, they will destroy the natural environment. There is a three-word mantra in the field of environmental engineering 'RRR',—Reduce, Recycle and Reuse. This principle could be applied for controlling or removing marine contaminants from water. In this regard, readers in the market will find delightful and existing trends in the field of marine pollutants removal through this book, which is written by Dr. P. Senthil Kumar. He has more than two decades of experience in the field of environment pollution remediation through various technologies like adsorption, microbiological innovations, electrochemical reduction and many more. He is addressing the environmental issue of disturbing the ecology and its related properties. The author involved himself in writing this book by collecting lots of articles and technologies from various private sectors and published papers. His address to the nation regarding marine pollutant removal through novel strategies will help readers across the world in gathering more knowledge in removal technologies and the type of pollutants in marine water. These pollutants can be further addressed by readers in various sectors like providing many more novel removal properties.

This book will describe concepts on various pollutants that disturb marine water and their interactions with marine ecology. Some of the recent strategies for the removal of these pollutants are discussed in an elaborate way for easy understanding. Also recent issues that are faced by the world in controlling marine pollutions are even highlighted with case studies. Some of the pollutants like microplastics, plastic debris and algal pollutions are addressed in an enlarged and neat manner with more scientific concepts. On the whole, the book finds a clear path to address marine pollution and its removal, offering the readers a simple portrayal of clear and clean ideas.

Preface

Purpose

The material in the book is appropriate for beginners to learn about the strategies used in treating removing/clearing marine pollutants from ocean. This book is appropriate for self-study and can be used for reference study, as well for the people working in this field. Many of the treatment methodologies are discussed in detail with appropriate points. Current methodologies that are adopted all over the world for cleaning marine debris are discussed and highlighted with simplicity.

Background

It is not mandatory to have knowledge of marine engineering. The concepts of pollutants and marine life are sufficient to use this book. This book provides clear knowledge on marine biota and its interactions with pollutants from various sources. It can be considered to be a mini review identifying the importance of marine water, impacts on marine biota, and the purpose of cleaning up the debris.

Organization

Before beginning any chapter, an abstract is provided that covers the concepts that are to be discussed in detail inside the chapter. The contents that are listed highlight the subdivision that is discussed inside the chapters. The concepts are highlighted with references from where the point is chosen and discussed. Each chapter ends with conclusions that provide short points of what has been discussed within. The technologies that are discussed are already known by the users which allow them to be enhanced in a better way.

Concept

The main concept that this book covers is the treatment technologies that are being used worldwide in order to remove marine debris in a safe way. Also the book shows some of the impacts generated by pollutants. It shows that major pollutants are entering the marine matrix and their toxicity levels. Some of the recent innovations in the marine clean-up are highlighted in Chapters 7–9.

Acknowledgement

I have received immense help from several sources during the preparation of this book and gratefully acknowledge various societies, journals, associations and several authors for the reproduction of salient features that are included throughout the text.

I wish to record my gratitude to all my scholars, particularly Ms. G. Prasannamedha, full time research scholar, in shouldering the domestic responsibilities and helping me to concentrate on the compilation of source material for the preparation of the text.

Finally, I also express gratitude to my parent institution, SSN College of Engineering, my students, friends, family members, colleagues and the publisher for their kind encouragement, cooperation and timely help extended during the preparation of the text.

Last but not the least, I express my soulful gratitude to almighty for giving me a chance in drafting this book and compiling on time.

I would like to dedicate this book to my daughter Ms. S. Vishrutha.

P. Senthil Kumar

CHAPTER ONE

Introduction to marine biology

1.1 Introduction

About 75% of the Earth's surface is covered with seawater, which is the major source of food, energy and mineral resources. This ocean water carries living organisms, hence it is called an aquatic ecosystem. The ocean helps to control the global climate that further depends on environmental factors like temperature, monsoon, human activities, etc. The aquatic ecosystem is classified into two types: freshwater ecosystem and marine water ecosystem. The marine ecosystem includes habitats of open seas, coastal zones, salt marshes and wetlands along shores and river mouths. Also it includes estuaries, tidal inlets and the foreshore ecosystem [1].

1.1.1 Nature of seawater

Seawater is a different solution that has water and chemical compounds as its constituents. When any external factor comes into contact with seawater, a degree of solubilization takes place. Water, which acts as a solvent medium, is an universal solution in which diffusion/transport of ions takes place. On the other hand, the chemical constituents of seawater include dissolved and undissolved substances. Dissolved substances include salts, organic compounds and gases. Undissolved substances include gas bubbles, inorganic and organic solids. The chemistry of seawater reveals the presence of six ions like Cl^-, Na^+, SO^{2-}, Mg^{2+}, Ca^{2+} and K^+. These ions are categorized as major constituents that contribute nearly 99.5% in seawater, whereas iron falls under the category of minor constituents [2].

1.1.2 Categories of marine ecosystem

The marine ecosystem is characterized by two components: biotic and abiotic components. Biotic components are living organisms like parasites, predators, competitors and other species. Abiotic components are temperature, salinity, turbulence, density, sunlight and concentration of nutrients. All these components are affected by various factors like quality of

Modern Treatment Strategies for Marine Pollution.
DOI: https://doi.org/10.1016/B978-0-12-822279-9.00008-7

seawater, buoyancy, gravity, temperature, density, penetration of light, water turbulence and hydrostatic pressure [1]. In general biotic factors include plants, animals, fungi, algae and bacteria. Abiotic factors include sunlight, temperature, moisture, wind, soil type and nutrient availability.

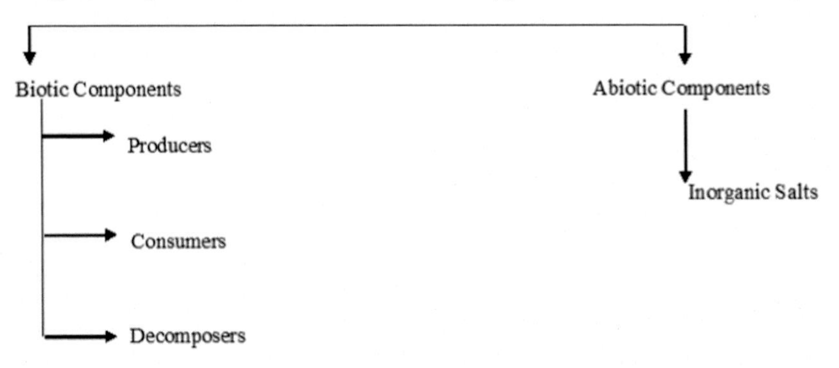

1.1.3 Presence of biotic and abiotic components in marine biota

The marine ecosystem provides information about marine organisms and their habitats which provide knowledge about the effects of global changes on marine ecology. This ecological study provides a better understanding of the interactions between organisms and environmental factors. This complex interaction provides a route for various external factors that are involved in the marine ecosystem. They are ocean temperature, dramatic changes in weather patterns, ocean acidification, melting of glaciers and pollution. There are a few factors that affect life in the ocean. Some of them are discussed below [3].

1.1.3.1 Temperature

Temperature is one of the most important factors that affects the rate of biological processes in ocean. In general the difference in temperature affects weather globally. It is known that increased water temperature results in weakened currents and less rainfall, which further affects physio-chemical and biological conditions of the ocean [3].

1.1.3.2 Nutrient concentration

Temperature change in the ocean affects the nutrient concentration, which in turn affects the biological growth of biota within ocean water.

This also affects the primary production of fish stocks and in turn the effects of fish stock production affect global yield and supply [3].

1.1.3.3 Influence of CO_2 levels

The influence of global industrialization affects the CO_2 level in the atmosphere. An increased level of CO_2 lowers ocean pH, producing acidification in ocean water. There are certain organisms that are affected by reduced pH, such as corals, bivalves and calcareous plankton, which in turn induces negative effects in the food web [3].

1.1.3.4 Melting of ice

Melting of ice causes a rise in sea level, freshening of seawater and reduction in the speed of water currents. A rise in sea level affects human habitation, whereas freshening of seawater affects all life forms due to alterations in the global climate [3].

1.1.3.5 Miscellaneous factors

There are other factors like salinity, UV, hypoxia and pollution that affect marine organisms in coastal areas. These factors show reduced resistance in biota like coral reefs. Also pollution induces toxicity to marine biota in aquatic systems [3].

1.1.4 Nature of biota in seawater

In the marine ecosystem the distribution of biological organisms is restricted by environmental parameters like pH, salinity, depth, temperature, pressure, current, nutrients and sediment properties. Classification of marine biology is based on sea bottom, depth, light and specific relationship to land.

The ocean has two broad environments: benthic and pelagic.

- Benthic—also called the bottom of ocean; it has greater variability.
- Pelagic—termed as a near water environment, especially water close to land that is affected by runoff and the influence of tides.

The benthic zone is subdivided into different zones, namely intertidal or littoral zone, supralittoral zone, sublittoral zone, bathyal zone, abyssal zone and hadal zone.

- Intertidal or littoral zone—the region of high-tide mark to low-tide mark
- Supralittoral zone—the beach or shore above high-tide lines influenced by ocean activities

- Sublittoral zone—away from land, that is, low-tide mark to edge of the continental shelf
- Bathyal zone—slopes and rises of the ocean floor
- Abyssal zone—the region of ocean floor plains
- Hadal zone—the region of deep trenches in ocean

The pelagic zone has two main subdivisions: neritic zone and oceanic zone.

- Neritic zone—starts at the edge of the low-tide mark and extends to the edge of the continental shelf
- Oceanic zone—the region between continental shelves

The oceanic zone is further subdivided into four types based on depth. They are epipelagic (0–200 m), mesopelagic (200–1000 m), bathypelagic (1000–3800 m) and abyssopelagic (greater than 3800 m) zone [2].

Organisms living in marine biota can be classified into planktons and nektons surviving in the pelagic environment, whereas benthos lives in the benthic environment. Some of the benthic producers are microalgae, macroalgae, sea grass, etc. Single/multicelled planktons of the photic zone are pelagic producers. Fig. 1.1 illustartes the zones that are commonly found in ocean.

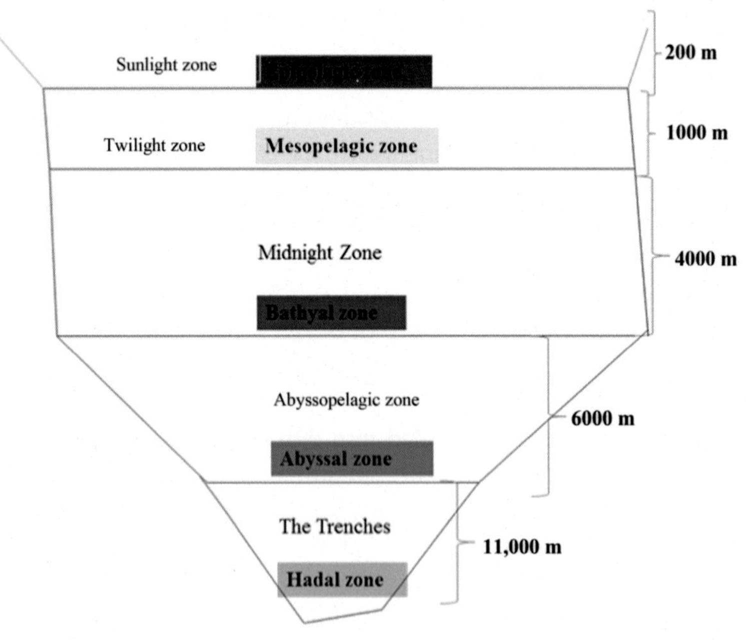

Figure 1.1 Oceanic zones.

1.1.5 Environmental factors influencing the marine ecosystem

Every species lives within an environment whose tolerance limit depends on the environmental factors like salinity, temperature and hydrodynamic conditions. In the marine environment there is a strong relationship between the abiotic habitat and the biological community it supports. One of the important factors that defines the occurrence of species is the type of substratum, which can be divided into rock and sediments. Some important factors that affect the marine community are listed below [4].

- Substratum
- Exposure to wave action
- Strength of tidal currents
- Salinity
- Temperature
- Topography
- Geology
- Oxygenation
- Wave surge
- Scour, turbidity and siltation
- Shading
- Organic carbon
- Hydrographic regime

1.1.6 Anthropogenic factors influencing marine ecosystem

Physical disturbance, like tourism, fisheries and dredging, affects rocky habitats of the marine ecosystem. Other activities like aggregate extraction and benthic fisheries activities may affect the sediment habitats of the marine environment. Severe pollution may reduce species richness and encourage higher densities of opportunist species [4].

1.1.7 Biotic and abiotic interaction

The interaction of the physical and biological process is important in visualizing communities in the marine environment. It is common that light, nutrients, temperature and feeding behaviour are the universal factors that affect the interaction. There are other factors that control the interaction between biotic and abiotic components. These factors are categorized as physical and biological processes.

1.1.7.1 Physical processes

- Climate, which is due to the coupling of incoming solar radiation and processes within the atmosphere, lithosphere, biosphere and cryosphere. Any alteration in any single mechanism affects the other components.
- Water currents, where, in accordance with water motion, horizontal movement is the most predominant, whereas the vertical direction is restricted due to a strong density gradient.
- Temperature may influence the population and community responses to climate change. It has a direct control on productivity.
- Hydrodynamic processes in the ocean may control production in the ocean by governing irradiance and nutrients. Saturating nutrient level may lower irradiance which is followed by low productivity. It is possible to have a limited irradiance level with low nutrients in water. The physiological response of communities to variation in light and nutrients occurs in a short span of time.
- Irradiance occurs in two forms. Most common occurence is quantified amount of irradiance at the surface of ocean are low that commonly occurs during winter and over ice. The other is depth-time integrated photosynthesis is minimum than depth-time integrated respiratory demand. This condition occurs during vertical mixing.
- Vertical mixing is a process in which energy is transferred across depth via turbulence diffusion. It is responsible for the input of nutrients from water into the deep sea that has direct impact on marine communities.
- One of the major interactions is diffusion, which can be categorized into two different types: molecular and eddy diffusion. Molecular diffusion is where molecules are distributed evenly. Eddy or turbulent diffusion takes place in such a manner that vertical gradients of heat, salt, nutrients and particulate matter are reduced or eliminated [5].

1.1.7.2 Biological processes

Biogeochemical cycles are much related to life and the Earth's environment through the flow of energy and matter. The presence of external matter like chemicals and pollutants affects the growth of marine biology. Factors like vertical mixing and spatial distribution have impacts on marine biota. The presence of nutrients also affects the life span and cycle of marine biology [5].

Higher nutrient and lower dissolved organic carbon concentrations, distinct microbial community compositions among habitats and assemblages of zooplankton that exhibit migration behaviour on marine biota are responsible for transport of pollutants within marine body. Physical and biological processes are responsible for the high retention of organic matter within marine biota that further affects the marine ecosystem [6].

1.1.8 Different types of interaction in marine biota

1.1.8.1 Interactions at physicochemical levels

Physical processes like warming and freshening in polar oceans have an impact. Warming and freshening increase density stratification, which further reduces mixing of high-oxygen nutrient-poor surface waters with deeper waters that have minimal oxygen. It can also alter concentrations of gases like carbon dioxide and oxygen. There are also some chemical interactions between chemical species and physical parameters, for example the influence of ammonia on nutrient levels [7].

1.1.8.2 Interaction at organism level

It denotes the growth rate and photosynthetic competence. It is categorized into three types, namely, independent nutrient colimitation, biochemical substitution colimitation, and biochemically-dependent colimitation [7].

1.1.8.3 Interaction at ecosystem level

There are a wide range of interactive effects that act through ecosystems. They are (1) variable sensitivities of organisms to drivers; (2) effects that directly alter the characteristics of the predator—prey relationship; (3) effects that indirectly alter the characteristics of the predator—prey relationship; (4) the collective effect of (1)—(3) across many facets of ecosystem functioning [7]. Fig. 1.2 denotes the interaction of marine biota with the ecosystem.

1.1.9 Blueprint on marine pollution

Basically land-based sources like agricultural runoff, discharge of nutrients, pesticides, oil spills from accidental discharge, untreated sewage and effluents [8] are pollutants that are threatening the ocean and its components in many forms. For example the discharge of sewage or effluents in seawater may increase the incidence of low oxygen (hypoxic), thereby leading to the formation of dead zones where marine species cannot survive.

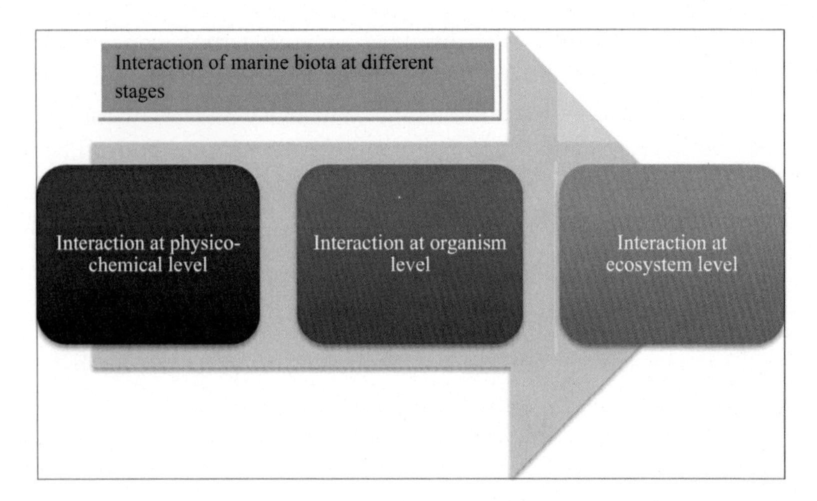

Figure 1.2 Interaction of marine biota with the ecosystem.

There are 500 dead zones with a total global surface area of over 245,000 km^2, roughly equivalent to that of the United Kingdom. Additionally, an increase in nitrogen content may increase the proliferation of seaweeds and microorganisms, or algal blooms which kill fish, thus contaminating sea food with toxins and changing the ecosystem. Plastic litter that is found floating in ocean water may resist light entering the deep ocean leading to breakdown of plastics to microparticles that are mistaken as food and consumed by marine biotas. Oil spills are also another major pollution that results from accidents occurring in the ocean during the transportation or extraction of oils. The best example is the Gulf of Mexico Deepwater Horizon oil spill, which had a devastating effect on the entire marine ecosystem, as well as the populations that depended on the marine areas for their livelihoods. Smaller oil spills happen every day, due to drilling incidents or leaking motors, and cause deaths [8].

These pollutants affect marine life and in turn may affect human life and habitats. For example due to these pollutants some beaches are closed as they were found to be unfit for bathing and cleaning purposes, due to plastics and microparticles found floating at them. These particles are consumed by fish and shellfish and they may end up on our plate and cause food poisoning [8]. In addition, these pollutants cause millions of dollars to be spent by industries due to various harms, such as:

- Discarded plastics gets caught in boat propelers and cooling intake, damaging the engines.
- Ghost fishing that occurs when discarded fishing nets entangle marine life indiscriminately, thus reducing fishers' revenues from lost catch.
- Extinction of fish species would lead to loss of biodiversity [8].

1.1.9.1 Facts and figures on marine pollution

- Land-based sources (such as agricultural runoff, discharge of nutrients and pesticides and untreated sewage including plastics) contribute around 80% of marine pollution, globally.
- Agricultural practices, coastal tourism, port and harbour developments, damming of rivers, urban development and construction, mining, fisheries, aquaculture and manufacturing are all sources of the marine pollution that is threatening coastal and marine habitats.
- The United Nations Environment Programme estimated in 2006 that every square mile of ocean contains 46,000 pieces of floating plastic.
- Plastics contribute as a carbon source for marine life and its ecosystem
- Once discarded, plastics are weathered and eroded into small fragments called microplastics that are found on every beaches everywhere.
- Plastic debris causes the deaths of more than a million seabirds every year, as well as more than 100,000 marine mammals.
- Plastic materials and other litter can become concentrated in certain areas called gyres as a result of marine pollution gathered by oceanic currents. There are now five gyres in our oceans [9].

1.2 Conclusion

It can be concluded that different biota in the marine ecosystem are responsible for interactions within marine biology. There are numerous factors that are responsible for the migration of biotic and abiotic components within marine bodies, such as, nutrient, temperature, light and hydrodynamic processes. There are different types of interaction levels involved in marine ecology. They are interaction at physicochemical level, interaction at organism level and interaction at ecosystem level. All these interactions favour the growth of marine biota.

References

[1] Balasubramanian, A. Aquatic ecosystems – marine types 2011. doi: 10.13140/RG.2.2.29494.09289.

[2] Florian M-LE. The underwater environment. Conservation of Marine Archaeological Objects 1987;1–20. Available from: http://dx.doi.org/10.1016/b978-0-408-10668-9.50007-1.

[3] Turkoglu M, Onal U, Ismen A. Introductory chapter: marine ecology—biotic and abiotic interactions Marine ecology – biotic and abiotic interactions. IntechOpen; 2018Available from. Available from: https://www.intechopen.com/books/marine-ecology-biotic-and-abioticinteractions/introductory-chapter-marine-ecology-biotic-and-abiotic-interactions.

[4] Connor DW, Allen JH, Golding N, Lieberknecht LM, Northen KO, Reker JB. The National Marine Habitat Classification for Britain and Ireland. Version 03.02. Introductory Text. Peterborough: Joint Nature Conservation Committee; 2003. Available from: www.jncc.gov.uk/marinehabitatclassification.

[5] Daly K, Smith W. Physical-biological interactions influencing marine plankton production. Annu Rev Ecol Syst 1993;24:555–85. Available from: www.jstor.org/stable/2097190 Retrieved March 20, 2020.

[6] Leichter JJ, Alldredge AL, Bernardi G, Brooks AJ, Carlson CA, Carpenter RC, Edmunds PJ, Fewings MR, Hanson KM, Hench JL, et al. Biological and physical interactions on a tropical island coral reef: transport and retention processes on Moorea, French Polynesia. Oceanography 2013;26(3):52–63. Available from: https://doi.org/10.5670/oceanog.2013.45.

[7] Boyd PW, Brown CJ. Modes of interactions between environmental drivers and marine biota. Front Mar Sci 2015;2:9. Available from: https://doi.org/10.3389/fmars.2015.00009.

[8] www.unesco.org. <http://www.unesco.org/new/en/natural-sciences/ioc-oceans/focus-areas/rio-20-ocean/blueprint-for-the-future-we-want/marine-pollution/>.

[9] www.unesco.org. <http://www.unesco.org/new/en/natural-sciences/ioc-oceans/focus-areas/rio-20-ocean/blueprint-for-the-future-we-want/marine-pollution/facts-and-figures-on-marine-pollution/>.

CHAPTER TWO

Biological and chemical impacts on marine biology

P. Senthil Kumar and G. Prasannamedha
Department of Chemical Engineering, Sri Sivasubramaniya Nadar College of Engineering, Kalavakkam, India

2.1 Introduction

Marine pollution is defined as the introduction of substances from humans into the marine environment resulting in such harmful effects as harm to living resources, hazards to human health, hindrance to marine activities including fishing, impairment of quality for use of seawater and the reduction of facilities. Marine water is subjected to contaminants coming from various sources that are commonly found in the surroundings. These contaminants change the characteristics of ocean and coastal zones, thereby affecting the biodiversity of the marine ecosystem, the quality of ocean water and productivity from marine ecology. Humans around the world discharge sewage, industrial waste, chemical waste and radioactive materials into the ocean. These pollutants either sink into the ocean depths or float. They are consumed by small marine organisms, thereby affecting the food chain. There are other pollutants like light, noise, chemical and plastic pollution that disrupt the marine ecosystem. This chapter briefly discusses the various types of pollutants, for example plastics, oil spills, organic contaminants and the presence of nitrogen and phosphorous contents in water, along with their impacts associated with marine biota. Also the chapter gives precise details on how the marine ecosystem is affected due to the release of these pollutants without proper precautionary measures.

2.2 Categories of pollutants in marine environment

There are three main sources of pollutants entering the marine environment. They are discharge as effluents, disposal of solid waste into water

Modern Treatment Strategies for Marine Pollution.
DOI: https://doi.org/10.1016/B978-0-12-822279-9.00006-3

and runoff via rivers. All these means of discharge greatly depend on the substance and situation in the environment [1]. Pollutants are categorized as follows.

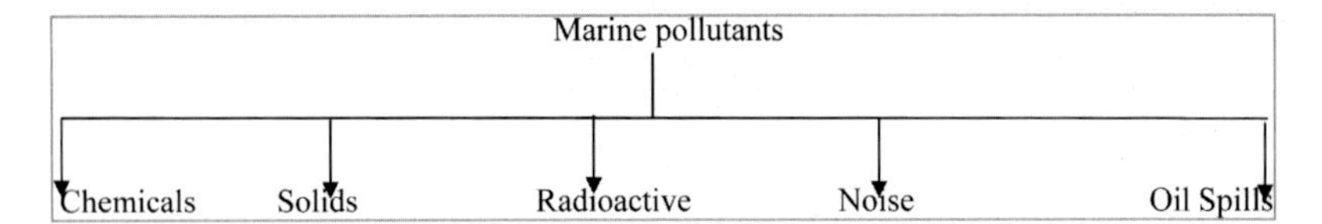

The following will explain the types of each pollutant and their impacts on the marine ecosystem [1]. Fig. 2.1 will explain the categories of marine pollutants.

2.2.1 Chemical compounds

The production and use of chemicals creates emissions to marine water in day-to-day life. The categories of sources of chemical emissions are [1]:

- Intentional dissemination of chemical products, for example pesticides
- Unintentional dissemination of chemical products, for example sewage leakage, pharmaceuticals, leakage, PCB, etc.
- Unintentional dissemination of chemical by-products, for example dioxins

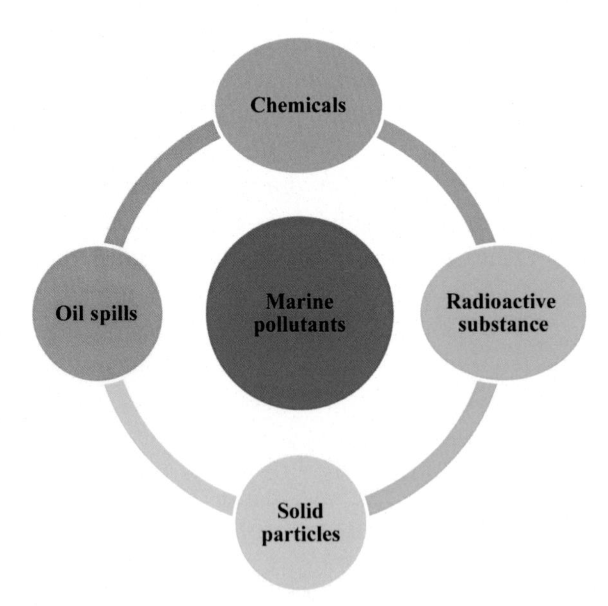

Figure 2.1 Categories of marine pollutants in marine ecosystem.

The main sources of chemical emissions are landfills, industries, sewage discharge and PCB industries. Examples of the chemical contaminants are metals, PCBs, flame retardants, plastic additives and other organic compounds. The main mechanisms for the survival of chemical compounds are bioavailability, bioaccumulation and toxicity. Bioavailability is the persistence of pollutants in an environment that determines the rate of removal. The removal can either be biological or physicochemical degradation. Bioaccumulation is the process in which substances accumulate in organisms at a higher concentration than in the surrounding environment. The main properties that define the accumulation of compounds in biota are hydrophilic and hydrophobic properties. Toxic effects are induced due to the direct exposure of organisms to chemicals in water, air and soil [1].

2.2.2 Oil spills

Petroleum or crude oil is the major oil contamination that is witnessed in marine water. Crude oil consists of a complex mixture of hydrocarbons with varying molecular weight. They also have nitrogen and sulfur moieties in small quantities. Hydrocarbons are alkanes and cycloalkanes with high boiling points [1].

2.2.3 Solid substances

These are substances varying in size that are deliberately or unintentionally released into the marine environment. They are categorized into three main types: marine debris, sedimentation and main tailings. According to the European Union, marine debris is defined as any form of manufactured or processed material discarded into marine water; examples are plastics, woods, metals, glass, paper, rubber and clothing. Increased sedimentation leads to an increased amount of turbidity in water, thereby threatening marine habitats [1].

2.2.4 Radioactive waste

These are the nuclear wastes that are most commonly seen in marine water. They are ^{3}H, ^{14}C, ^{90}Sr, ^{99}Tc, ^{137}Cs, ^{210}Po, $^{238,239,240,241}Pu$ and ^{241}Am. They may be in solid or liquid waste form and they may be available in soluble or insoluble form. They adhere to the particles of the marine ecosystem and induce serious issues to the biota [1].

2.3 Effect of plastic debris

Plastics are an integral part of daily life. It is estimated that around 5% of plastics are transported through wastewater, inland waterways, winds and tides, reaching the marine environment as plastic debris or litters. Around 80% of plastic waste is found in the marine habitat; plastic bags are most common. Contamination of the marine habitat by plastic bags leads to negative socioeconomic consequences. The coastal ecosystem provides the majority of services like the production of food, stabilization of shorelines, removal of pollutants and nutrient turnover. Sedimentary habitats support recycling nutrients that are important for benthic and pelagic food webs through myriad biogeochemical processes. It is found that plastic debris might decrease the biomass of microphytobenthos through blocking of the sunlight which is required for food productions. It is also found that plastic debris on the surface of marine water interferes in nutrient exchange processes at the sediment water interface, thereby forming anoxic conditions and decreasing the quantity of infaunal organisms [2].

Plastics are known to be a persistent material that can survive for hundreds of years in the marine environment. They are categorized as solid waste substances found in marine water bodies. Plastics being buoyant in seawater they sink in the sea beds thereby leading to fouling of marine organisms. Plastics break down into smaller fragments leading to microplastics which are known to be affecting the food web through ingestion by small marine organisms [1].

Generally plastics are divided into three types. They are plastics, microplastics and nanoplastics. Some of the common impacts of plastics debris are:

- Ingestion of plastic pieces by marine organisms
- Toxic to the marine matrix
- Tendency to adsorb chemical pollutants like PCBs that are present at trace levels
- Damage to the ecosystem through toxicity effects
- Health hazards like decreased immunity, fertility and sexual disorders can be witnessed in marine organisms [1,2]

2.3.1 Classes of plastic debris

Fig. 2.2 shows different classes of plastics based on size.

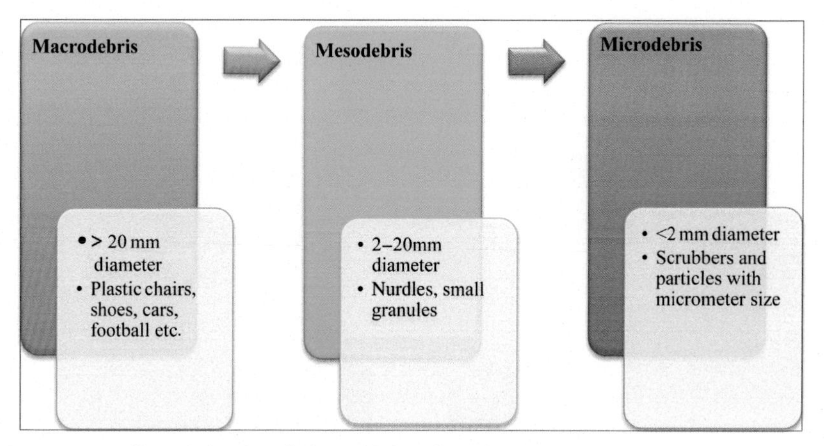

Figure 2.2 Different classes of plastic debris found in the ocean.

2.3.2 Sources of plastics in marine environment

There are numerous sources for the introduction of plastics into the marine environment. They are broadly categorized into two types: ocean-based debris and land-based debris. Some of them are described schematically below [3].

2.3.3 Sources of ocean-based debris

Nearly 5.6 million tons per year of ocean debris enter into the marine environment. They are solid waste that are discarded in ocean [3]. Fig. 2.3 shows different sources of debris in ocean.

2.3.4 Sources of land-based emission

About 12% of total solid waste entering the ocean are from land-based sources. Mostly high populated countries emit waste from land sources. Some of the sources are

- Municipal landfills
- Solid household waste
- Industrial facilities
- Tourism
- Discharge from treated and untreated sewage
- Transport of waste by rivers from landfills or any other sources of debris along river and waterways system [3]

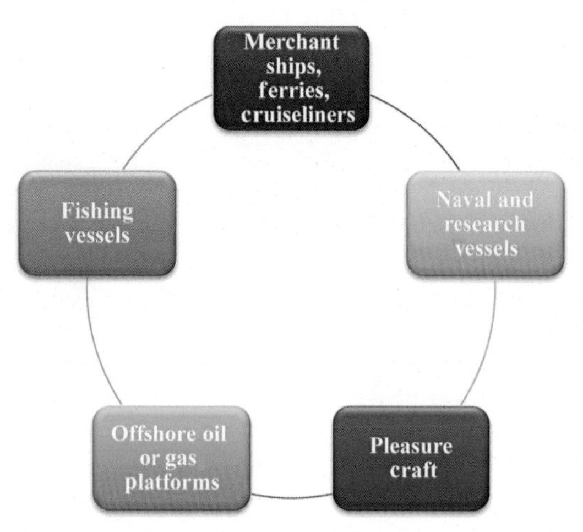

Figure 2.3 Sources of debris in ocean.

2.3.5 Degradation of plastic debris in marine environment

The degradation of polymeric materials in the environment occurs through the following reactions [3]

- Photooxidative degradation
- Thermal oxidation
- Biodegradation
- Hydrolysis

Generally plastics are broken down by solar radiation, which primarily depends on the wavelength and amount of radiation a polymer can absorb and the strength of chemical bonds within the polymer. There are two main mechanisms through which solar radiation degradation takes place [3].

- The reaction is initiated by the photolysis of chromophores as a result of absorbing UV radiation, producing hydroxy radicals.
- A photooxidative chain reaction is initiated by energy absorbed by impurities.
- The radicals created react with oxygen and polymer to produce cross-link bonds.
- Thereby the polymer loses its ability and strength [3]. Fig. 2.4 illustrate the various ways of degrading plastics in marine environment.

2.3.6 Impacts of plastic debris in the marine environment

Plastics are lethal when introduced into the environment. They cause damage to the marine organisms upon ingestion. Also they damage

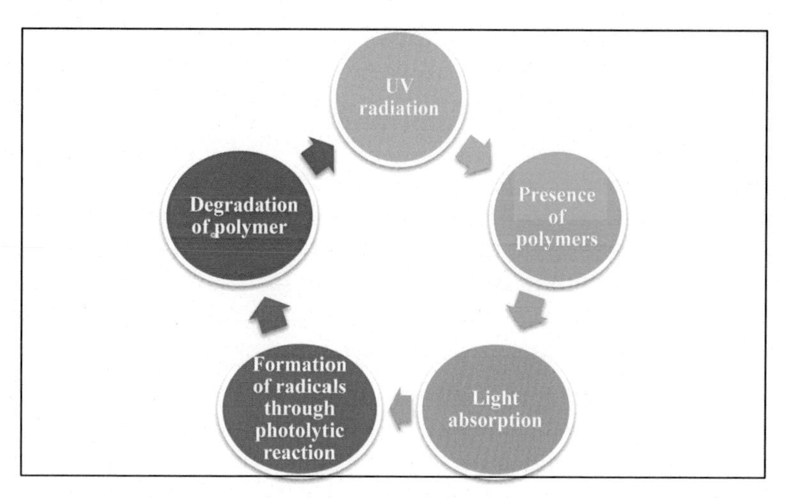

Figure 2.4 Various ways for the degradation of plastics in marine water.

marine industries by entangling propellers and blocking cooling systems. Some of the impacts are discussed below [3].

2.3.6.1 Mechanical impacts

Nearly 256 marine species have suffered from entanglement, which leads to the death of the organism. Entanglement can cause death by drowning, suffocation, strangulation, or starvation. Many coastal and marine species like seals, whales, fish and sea turtles are affected [3].

2.3.6.2 Ingestion

The ingestion of plastics in the marine environment is primarily due to being mistaken for food. Sometimes the consumed plastics debris, like micro debris or meso debris, passes through the gut without causing harm. But in most cases plastics get trapped in the stomach, digestive tract, or throat, leading to starvation [3]. Ingested plastics cause a reduction in stomach capacities, hinder growth, causes stomach injuries and create intestinal blockages. The amount of plastic ingested by various species may be an indicator of the accumulation of plastics in a chosen area. It is known that some of the lakes are known for discarding plastics from industries to ocean through cyclonic surface currents due to change in wind velocity and temperature [4].

2.3.6.3 Chemical impacts

Plastics by themselves are toxic to the environment. They are prepared in combination with a variety of chemicals, for example additives. These chemicals are not often chemically bonded, therefore they are able to leach from the plastics and react with the environment. They penetrate the cell membrane, interact biochemically and induce toxic effects. A few chemicals get adsorbed on the plastics that turn out to be stronger pollutants. Some of the chemicals that are toxic in the marine environment are additives and hydrophobic chemicals are adsorbed from the surrounding water. The toxic chemicals include phthalates, bisphenol A, brominated flame retardants. These chemicals cause neurotoxicological effects on mammals and aquatic organisms. The adsorption of chemicals on plastics reduces the transport and diffusion of contaminants. Hydrophobic organic contaminants have greater affinity for polyethylene, polypropylene and polyvinyl chloride. By adsorbing on plastics these chemicals not only get transported but their persistence in the environment increases [3].

2.4 Nitrogen and phosphorous imbalance

In the last few decades marine eutrophication is a massive problem that is faced globally. Marine eutrophication is mainly due to the presence of nitrogen and phosphorous in water. Its presence in marine water leads to hypoxia and anoxia, reduced water quality, habitat degradation, loss of food web structure, loss of biodiversity and growth of harmful algal blooms. At normal levels nitrogen and phosphorous are needed to support aquatic plant growth by promoting protein synthesis and DNA and RNA synthesis. Nitrogen is available in biological forms like nitrate, nitrite and ammonium in soil and water. Phosphorous is available in the form of organic and inorganic compounds. In an inorganic form they are crystalline and amorphous, whereas the organic form remains as a mixture of phosphate diesters, monoesters and polyphosphates. There are two cycles that are available in the environment for fixing nutrients for survival: the nitrogen cycle and phosphorous cycle. The nitrogen cycle contains the gaseous, particulate and dissolved forms of nitrogen and the phosphorous cycle contains particulate and nongaseous forms of phosphorous. This clearly shows that nitrogen can escape in gaseous form, whereas phosphorous gets trapped in the marine environment. Agricultural fertilizers, indigenous soil phosphorous, atmospheric deposition and anthropogenic phosphorous are the major source for nitrogen and phosphorous in the marine environment [5].

Eutrophication leads to the excessive growth of plants, the growth of harmful algal blooms, increased frequency of anoxic conditions and increased death of fish in marine biota. Some of the issues due to eutrophication are discussed below. They are [5]

- Ocean acidification
- Dead zones
- Human/animal health
- Tourism

2.4.1 Ocean acidification

Anoxic and hypoxic water are associated with the variation carbon dioxide, causing ocean acidification that alters the ocean chemistry and interferes with the emissions of carbon dioxide to the atmosphere. It affects calcifying organisms like mollusks and crustaceans [5].

2.4.2 Dead zones

Hypoxic and anoxic conditions lead to the formation of dead zones by destroying the organisms and reducing biodiversity. This alters the accumulation of particulate organic matter, changes microbial biochemical pathways and leads to the consumption of dissolved oxygen in bottom waters, resulting in the death of fish and other marine biota [5].

2.4.3 Human/animal health

The accumulation of algal blooms leads to an increased quantity of biomass, thereby affecting the microbial culture by producing toxins that are harmful to the surroundings [5].

2.5 Evolution of oil spills/oil dispersant and its impacts

Environmental pollution due to petroleum-leaked products has been seriously addressed over the past few decades. This is because hydrocarbons are toxic to aquatic and terrestrial ecosystems. These hydrocarbons include straight and branched chains, single or condensed rings and aromatic rings such as the monocyclic hydrocarbons. They can also be polycyclic aromatic hydrocarbons (PAH). They are toxic to the surroundings. This toxicity depends on various factors like oil composition, characteristics, weather condition, exposure routes and bioavailability. A major issue caused by oil spills is narcosis which is a reversible anaesthetic effect caused by the oil partitioning into the cell membrane and nervous tissue. It causes dysfunction of the central nervous system. It can induce serious mortality if the threshold level is reached. When oils are ingested by marine organisms they affect the liver where enzymes activate PAHs to become more toxic and reactive products. PAHs have oxidative and carcinogenic properties due to their ability to attack and bind to DNA and proteins of the ingested organisms [6].

Oil dispersants are chemical ingredients that are used after oil spills in marine water bodies. They break oil slicks and increase the rate of biodegradation. The organisms that have very little protective shell or have external tissues are prone to be sensitive to exposure to oil dispersants [6]. The upcoming section will discuss the fate and a few impacts of oil spills in the marine environment.

2.5.1 Fate of oil spills in marine environment

Following are the fate of oil spills when spilled in marine water [6].

- Oil spilled in ocean moves through winds and water current and spreads out
- At the same time it undergoes physical and chemical changes
- Induces weathering effect—breakdown of oil spills and become heavier than water
- Formation of water-in-oil emulsion through which oil becomes more persistent
- A few processes, like dispersion can remove oil spills from marine water.

2.5.2 Impacts of oil spills in marine water bodies

The impact of an oil spill depends on its fate. Oil, once discharged in marine water, either gets dispersed in the top layer of water or remains on the surface of the water. If oil remains on the surface it may move toward a coastal region, thereby affecting invertebrates, mammals and birds. Some zooplanktons, such as copepods, euphausids and mysids, consume hydrocarbons directly, thereby leading to mortality or developmental and reproductive disorders. Benthic invertebrates may accumulate petroleum hydrocarbons from water, sediments and foods. This higher accumulation of oil is due to a lack of mixed function oxygenase that makes them unable to metabolize hydrocarbons to execretable primary metabolites. There are many biota that are affected by oil spills including coral reefs, marine fishes, sea birds and marine mammals [6]. It is found from research that oil dispersants are also toxic to marine life, including short and hard coral reefs, at early stages. This is because corals are very sensitive to dispersed oil and oil dispersants. Generally fishes do not accumulate and retain high concentrations of hydrocarbons because of highly developed hepatic mixed function oxygenase. But there is research that proves the impact of oil spills at early stages of fish development, for example melanosis, erratic swimming, reduced mobility and startled responses. Natural variation and the enormous series of factors that influence bird population statistics make it hard to measure the impact of oil spills on sea birds. There are two major ways to assess the impacts of oil spills on sea birds: (1) clean and rehabilitate birds and release them to wild; and (2) assess the impacts on the population of species that are affected [6].

Marine mammals like bottlenose dolphins, fin whales, humpback whales, right whales, sei whales, sperm whales, manatees, other cetaceans, sea otters, seals and other pinnipeds rely on their outer coats for buoyancy and warmth. Hence physical contact of these mammals with oil spills affects them. Additionally it is found that marine microbes respond to oils spills depending on oil composition, degree of weathering, temperature and nutrient concentration. The presence of hydrocarbons in the marine ecosystem may lead to portioning of microorganisms. Oil spills may lead to the formation of cyanobacterial mats in seashore areas. There is much evidence that oil spills have serious effects on marine biota. Many researchers have done real-time studies to analyze the effects of oil spills on marine life [6,7].

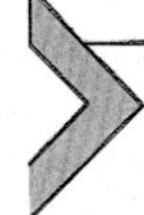

2.6 Organic contaminants interaction with marine biota

Organic contaminants include the presence of chemically active compounds that are discharged from industries. These may reach marine water through migration from source by water currents. These chemically active compounds include microplastics, leaching of chemical additives and persistent organic pollutants. They are ingested by green algae and may cause disruption to photosynthesis processes. Also they affect marine invertebrates and vertebrates by transferring as prey. Ingestion, retention and egestion are the major physical interactions of organic contaminants with marine biota. These processes can impair the nutritional health of organisms. They pave the way for a reduction in structural growth, maturity and reproduction in many marine biota. A few studies have reported that they affect intracellular metabolic and endocrine functioning processes. Chemically, they damage enzyme activity and gene expression and cause oxidative damage which lead to sublethal pathological responses in marine biology. Upon ingestion the chemical modification of organic compounds takes place within the organisms, thereby leading to an increase in retention time which may turn out to be toxic to them. Some organic contaminants are initiators, catalysts, solvents, stabilizers, plasticizers, flame retardants, pigments, fillers, nonylphenol and bisphenol A. Sorption is a mechanism through which organic contaminants persist [8].

2.7 Key problems in marine ecosystem

The marine ecosystem is affected because of increasing human activities and global warming effects and thus it faces a number of problems [9]. Some of the key problems in the marine ecosystem are:

- Discharge of emerging contaminants into seawater due to human activities
- Production of algal toxins or red tide toxins during algal blooms due to increased organic matter in water
- Acidification of marine surface waters
- Overfishing depleting the ecosystem
- Shipbreaking and recycling industries, along with oil exploration and transportation may have catastrophic effects on biodiversity [9]

These are some of the main problems that are increasingly affecting the world marine ecosystem as they are directly or indirectly linked with human population. Coastal areas are understandably suffering from the biggest impact and human activities have depleted >90% of formerly important species, destroyed >65% of seagrass and wetland habitats, degraded water quality and accelerated species invasions in diverse and productive estuaries and coastal seas. Insufficient attention has been paid by developing countries for the declining in marine ecological communities. Considering the importance of marine resources care must be taken toward the problems created by human activities on the marine ecosystem [9].

2.8 Ocean acidification and its impacts

It is the fact denoting increase in ocean pH. Seawater acidification is linked with atmospheric CO_2 that has connection with global warming. It is observed that average pH in surface ocean dropped by 0.1 units denotes that increase in $[H^+]$ reaches 30%. Seawater acidification is related to the pH value of the surface water layer which is substantially increased during summer stratification period or during algal bloom growth due to an increased content of water DOM and POM. A pH increase in surface stratified water is due to products of DOM and POM that are formed due to photodegradation and photosynthesis that consumes CO_2 [9]. Global warming is one reason for seawater acidification as it lengthens the summer stratification period, which increases the availability of CO_2, thereby enhancing the release of organic matter. This shows that acidification acts in a complex system where it might be difficult to completely distinguish which ecosystem effects can be purely attributed to a pH decrease [9].

2.8.1 Impacts of ocean acidification

Ocean acidification would decrease the saturation states of carbonate materials and change the calcification rate of some marine organisms. They affect shellfish, coral reefs and other marine organisms [9].

- Shellfish or marine calcifiers are sensitive to an increase in acidity/decrease in pH. It causes the dissolution of magnesium calcite, which is an important component of these organisms. It shows an impact on fertilization, sexual reproduction, cleavage, larval settlement, survival and growth, thus finally causing a substantial population decline [9].
- Coral reefs are sensitive to acidification. Acidification can dissolve reef carbonate; reduce the development of coral larvae into juvenile colonies; decrease growth rates of juvenile scleractinian corals; increase

sperm mortality; cause a decline in the early developmental stages (fertilization, sexual reproduction, metabolism, cleavage, larval settlement and reproductive stages); reduce algal symbiosis and postsettlement growth; delay the onset of calcification and alter crystal morphology and composition; increase juvenile mortality because of slower postsettlement growth; and reduce effective population size. The synergistic effects of elevated seawater temperature and of CO_2-driven ocean acidification are responsible for coral bleaching, reduction of primary productivity and for the decline in growth and calcification rates [9].

- For some marine invertebrates, the calcification of larval and juvenile or smaller individuals is more sensitive to acidification compared to larger individuals [9].
- Photosynthesis, calcification and nitrogen fixation of some coccolithophores, prokaryotes and cyanobacteria are either unmodified or increased or decreased in high-CO_2 water. Algal toxins production might increase due to increased ocean acidification [9].

2.8.2 Factors affecting ocean acidification

Some of the factors that affect ocean acidification are eutrophication caused by photosynthesis, respiration, temperature, light and nutrients. All these factors would modify the water alkalinity and variation of pH upon CO_2 dissolution. Additionally, global warming can substantially increase the surface water temperature, which can enhance the rate of photoinduced and microbial degradation of DOM and POM. Atmospheric acid deposition and acid rain may increase the pH values and geochemistry of surface water. Agricultural activities, through the oxidation of nitrogen fertilizers to nitrate, can further contribute to the decrease of seawater alkalinity [9].

2.9 Shipbreaking and recycling industries

Shipbreaking is the process of cutting and breaking apart old ships to recycle scrap metals, along with the simultaneous scrapping or disposal of expired or unused ships. It is an important source of hazardous contaminants along the coastal seashore, especially in the case of old oil tankers, bulk carriers, general cargo, container ships and passenger ships. This can produce three kinds of pollutants: solid, liquid and gaseous waste. These wastes are categorized as follows. Liquid waste is oils and oil products (engine oil, bilge oil, hydraulic and lubricant oils and grease); persistent organic pollutants including polychlorinated biphenyls (PCBs, used, e.g. in transformers); PAHs; ozone-depleting substances (ODSs) (e.g. CFCs and Halons); preservative coatings; organotins including monobutyltin (MBT), dibutyltin (DBT) and tributyltin (TBT); waste inorganic liquids (e.g. sulfuric acid); waste organic liquids; reusable organic liquids. Solid waste includes various types of asbestos; paint chips; heavy metals such as mercury (Hg), cadmium (Cd), lead (Pb), arsenic (As), chromium (Cr), copper (Cu), manganese (Mn), iron (Fe), zinc (Zn), nickel (Ni) and aluminium (Al); polyvinyl chloride (PVC); solid ozone-depleting substances (ODSs, e.g. polyurethane); solid PCB-contaminated wastes (e.g. capacitors and ballasts); plastic; sludge; glass; cuttings; ceramics. Gaseous waste is sulfur fumes; dioxins produced during burning of chlorine-containing products, such as PCBs and PVC [9].

2.9.1 Impacts of released pollutants from shipbreaking

As they carry lot of pollutants from various sources they are prone to have huge impacts on human and environmental zones. Some of the impacts are:

- Decline in fish communities
- Decline in primary production and zooplankton communities
- Decline in benthic invertebrates and their stock
- Adverse health problems for nearby workers in the environment after exposure to contaminated flora, fauna and other factors
- Unsuitable and harmful seawater for recreational purposes
- Toxic effects and population decline for marine birds, mammals, crustaceans, turtles and reptiles, through uptake of contaminated fish, polluted waters and other seafood
- Contribution to acid rain through atmospheric emissions
- Overall loss in flora and fauna surrounding the marine ecosystem [9]

2.9.2 Factors affecting shipbreaking pollutants

It is observed that most of pollutants are discharged into the marine environment without any pretreatment. Some of the reasons that serve as a factor for the causes of alarming increases in pollutants are [9]

- Lack of knowledge about the environmental impacts of these pollutants

- Lack of technology or recycling options to control the release of pollutants from shipbreaking
- Being unwilling to take remedial measures due to profit issues
- Lack of proper rules and regulation to control these pollutants [9]

Hence it is mandatory to make safe and environmentally sound yards for these pollutants to solve the pollution problems. Every country should follow sustainable development and follow certain obligations:

- Each shipbreaking yard should be conducted in an exclusive zone, from which pollutants could not be directly released into marine and terrestrial ecosystems.
- Recyclable unit should be incorporated.
- Awareness should be raised among workers and consumers [9]

2.10 Eco-cycle communication between pollutants and biota

Production, use and dissemination of specific pollutants are influenced by the development of national and international agreements, policies and priorities. Available scientific data and levels of coordination data can quantify the global impacts due to pollutants in the marine ecosystem. Marine pollution induces threats to food quality and thereby affects human health. The distribution of pollution impacts is heterogeneous at a global level. Pollution impacts undermine the resilience of the ecosystem to other stressors like oil spills, plastics, etc. [1].

There are responsible factors that show the effect of population on marine pollution. The increasing world population has an increasing demand for food, medicines, goods and habitats. All these issues are directly or indirectly associated with marine problems, such as overfishing, increasing emissions of pharmaceuticals and other ECs, plastic waste, oil exploration and transportation and algal blooms. Another major problem is the decline in the fish community. The fast depletion of fish stocks by overfishing and environmental deterioration could constitute an economic and an ecological problem, ruining fishing communities and seriously damaging the whole fishing-based supply chain. Release of pollutants to the marine environment is a serious issue to public health, as the food supply is related with marine biota [9].

2.11 Conclusion

Marine pollutants have a serious impact on marine life, as well as on the economic coastal activities and the communities that exploit the resources of the sea. Generally the toxicity levels of pollution depend on the physical and chemical condition of environment. Marine life can also be affected by cleanup operations or indirectly through the physical damage to the habitats in which plants and animals live. Communities that are threatened by marine pollutants have realized the risk and have therefore developed their own plans and policy issues to counteract the risk of marine contamination. Due to different anthropogenic activities a number of socioeconomic impacts have been predicted. Hence it is the duty of every individual to control the release of pollutants and thereby stop the damage to the marine ecosystem.

References

[1] Wilhelmsson D, Thompson RC, Holmström K, Lindén O, Eriksson-Hägg H. Marine pollution. Managing ocean environments in a changing climate. Elsevier; 2013. p. 127−69. Available from: http://doi.org/10.1016/b978-0-12-407668-6.00006-9.
[2] Green DS, Boots B, Blockley DJ, Rocha C, Thompson R. Impacts of discarded plastic bags on marine assemblages and ecosystem functioning. Environ Sci Technol 2015;49 (9):5380−9. Available from: https://doi.org/10.1021/acs.est.5b00277.
[3] Hammer J, Kraak MHS, Parsons JR. Plastics in the marine environment: the dark side of a modern gift. Rev Environ Contaminat Toxicol 2012;220:1−44. Available from: https://doi.org/10.1007/978-1-4614-3414-6_1.
[4] Sigler M. The effects of plastic pollution on aquatic wildlife: current situations and future solutions. Water Air Soil Pollut 2014;225(11). Available from: https://doi.org/10.1007/s11270-014-2184-6.
[5] Ngatia Lucy, Grace III Johnny M, Moriasi Daniel, Taylor Robert. Nitrogen and phosphorus eutrophication in marine ecosystems. In: Fouzia Houma Bachari, editor. Monitoring of marine pollution. IntechOpen; 2019. Available from: https://doi.org/10.5772/intechopen.81869.
[6] Saadoun IMK. Impact of oil spills on marine life. Emerging pollutants in the environment - current and further implications. IntechOpen; 2015. Available from: http://doi.org/10.5772/60455.
[7] Jernelöv A. The threats from oil spills: now, then, and in the future. AMBIO 2010;39 (5−6):353−66. Available from: https://doi.org/10.1007/s13280-010-0085-5.
[8] Prinz N, Korez Š. Understanding How Microplastics Affect Marine Biota on the Cellular Level Is Important for Assessing Ecosystem Function: A Review. In: Jungblut S, Liebich V, Bode-Dalby M, editors. The Oceans: Our Research, Our Future. YOUMARES 9, Cham: Springer; 2020. Available from: https://doi.org/10.1007/978-3-030-20389-4_6.
[9] Mostofa KMG, Liu C-Q, Vione D, Gao K, Ogawa H. Sources, factors, mechanisms and possible solutions to pollutants in marine ecosystems. Environ Pollut 2013;182:461−78. Available from: https://doi.org/10.1016/j.envpol.2013.08.

CHAPTER THREE

Detection and monitoring of marine pollution

3.1 Introduction

The ocean acts as a natural place for the accommodation of carbon dioxide and other greenhouse gases. The balance in the marine ecosystem is disturbed when it is contaminated by external agents like plastics, oil spills, toxic chemicals, radioactive waste and sewage. Also the marine environment is polluted due to the excessive discharge of nutrients from agriculture, fertilizers and pesticides. The release of such pollutants in marine water creates massive damage to marine biota from primary organisms to tertiary organism. Additionally pollutant like microplastics find their place in marine life due to the unwanted discharge of organic contaminants. In order to avoid the release of harmful pollutants into marine water it is necessary to reduce their release, therefore it is mandatory to map and monitor marine pollutants to ensure a sustainable marine ecosystem. This chapter describes various tools used for monitoring pollutants, like sensing, analytical methods and electrochemical methods in the marine environment.

3.2 Importance on identification of marine pollution

Marine pollution leads to the disturbance of the ocean environment and its biota, thereby affecting the environment and human health. All pollutants reaching the marine environment will induce some biological impacts like death, metabolic malfunction and genetic damage. Additionally there are some impacts like loss of marine biodiversity, changes in habitat and the food chain that affect productivity patterns. Basically external factors affecting marine biota are termed stress. Stress to marine life may be physical, chemical, or biological

Modern Treatment Strategies for Marine Pollution.
DOI: https://doi.org/10.1016/B978-0-12-822279-9.00012-9

factors. All the stresses are due to anthropogenic activities. Every activity performed by human is not environmentally neutral, thereby leading to certain consequences to the surroundings. Some of the novel stresses that are produced by humans are climatic changes, microplastics, noise, pharmaceuticals, bioaccumulation and biomagnification [1].

Climate change affects the interaction between natural resources and the global climate. It affects the ecosystem, for example, loss of habitat, invasive species, production of climate-induced toxicity susceptibility and toxicant induced climate susceptibility. The toxic effect of microplastics is due to additives and monomers. They can absorb hydrophobic contaminants from water due to the large area to volume ratio. Chronic exposure to noise leads to permanent disruption to sleep. Noise is a major pollutant recognized by the World Health Organization and its effects are well-known. Bioaccumulation and biomagnification are finding the presence of toxic contaminants in marine life. They are prone to damage the tissue cells and reproductive capacity and cause other biological disorders. They cause some issues like the loss of biodiversity as well. Hence considering all these impacts it is mandatory to find the pollutants that are present or isolated in the environment [1].

3.3 Factors monitored with respect to marine pollution

Some of the factors that are considered in removing pollutants that are found in the marine ecosystem are [2]:

- chemical analysis of toxic pollutants;
- biochemical and physiological exposure of organisms exposure;
- measure of community response like biomass and biodiversity; and
- toxicity assay of water and sediments.

The main objectives of monitoring marine pollution are [2]:

- *Compliance*: To ensure that the activities are carried out in accordance with the permission or rules and regulations.
- *Hypothesis testing*: To check the validation of assumptions and predictions that are made during the assessment of toxicity or the impact due to any pollutant.
- *Trend monitoring*: To identify and quantify large-scale changes in the marine environment as consequences of multiple activities being managed.

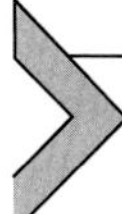

3.4 Remote sensing platform in monitoring marine pollution

3.4.1 Remote sensing

Remote sensing works by measuring the electromagnetic radiations that are returned from the surface of water to the sensor on an aircraft or satellite. The radiations that are reflected by the surface may vary in wavelength, which assists in detecting the variation in the surface that is to be monitored. There are some limitations to the use of electromagnetic radiations. They can differentiate only between surfaces that can be sensed. These radiations are very poor at penetrating into water. At the infrared wavelength they get absorbed at the surface of water and in the visible range they penetrate only a few meters [3]. Remote sensing with respect to marine pollution involves the following steps:

- Detection and tracking
- Damage assessment
- Abatement, prevention and construction of legislative control
- Monitoring [3]

3.4.2 Types of sensors

A variety of sensors have been used to image the marine environment. For example thermal sensors are used to image the sea surface temperature (SST). Sensors are classified as active and passive sensors. Passive sensors are sensors that do not require any external source to produce an output signal. Active sensors require an external power source to operate, transmit and detect energy at the same time [4]. Different types of sensors are discussed in Table 3.1.

Table 3.1 Types of sensor.

Type of sensor	Spectrum	Application	References
Passive	Visible and near infrared	Ocean colour, pigment concentration, bathymetry, surface slicks and suspended sediment concentration	[4]
	Thermal infrared	Sea surface temperature (SST)	[4]
	Microwave	SST, surface slicks, surface heat flux, sea state, ice, wind, wave condition	[4]
Active	-	Sea surface height, tides, surface roughness, interval waves, topography	[4]

3.4.3 Remote-sensing platform

There are several remote-sensing platforms available for monitoring water pollutants. They are categorized into two types: airborne and spaceborne sensors [5].

3.4.3.1 Airborne sensors

They fly at lower altitude. Therefore the data acquired will always have a higher level of detail. They are used for monitoring oil and chemical spills. There are four airborne sensors used for spill surveillance [5].

- Infrared/ultraviolet line scan (IR/UVLS)
- Side-looking airborne radar (SLAR)
- Microwave radiometer (MWR)
- Laser fluorosensor (LF)

3.4.3.2 Spaceborne sensors

They cover extensive remote areas to monitor water quality. The spatial coverage of sensors range from 10 to 100 km and temporal frequency is from hourly to weekly monitoring. The major applications are sea surface currents, oil spills, biogenic films and river plumes [5].

3.4.4 Remote sensing in principle and its role in ocean water monitoring

Reflection, transmittance and absorption of electromagnetic radiation mainly depend on concentration, types and the pollutant present in water. Total absorption is the sum of absorption by microalgae, nonalgal pigments (NAP), coloured dissolved organic matter (CDOM) and absorption by water. Light scattering in water denotes the presence of suspended solids (SS) in water bodies. Clear water has low reflectance in the visible spectrum and no reflectance in the near-infrared (NIR) region. A high concentration of chlorophyll shows reflectance in the green region, whereas reflectance in red and NIR regions shows the presence of SS in water [5].

SST is an important parameter to study the ocean circulation mechanism. Many satellite sensors such as moderate resolution imaging spectroradiometer (MODIS), visible infrared imaging radiometer suite (VIIRS), the advanced very high-resolution radiometer (AVHRR) and the sea and land surface temperature radiometer (SLSTR) measure the emitted thermal energy to determine SST. Fluorescence is another type of energy emitted by substances, e.g., algae, which can be detected using optical sensors [5].

3.4.5 Application of remote sensing

Table 3.2 shows some of the sensors that are used for various applications in the field of remote sensing. Fig. 3.1 depicts principle mechanism involved in remote sensing.

Table 3.2 Applications of sensors.

Parameter	Sensor	Application	References
Chlorophyll – Chl-a	CZCS – 1978	To deduce the growth of algal blooms in oceans by detecting the Chl-a concentration using satellite imaginary	[5]
	Synthetic aperture radar (SAR)	To detect large algal blooms in cloudy weather	
Oil spills	Airborne laser fluorosensor and microwave radiometers (MWRs) SAR equipped satellite ERS-2 SAR and RADARSAT-1 SAR	Measuring oil spills in surface of ocean during all day and all weather conditions	[5]
Marine plastics	Airborne sensors Sentinel-1A and 1B, TerraSAR-X	To identify the floating plastics in marine water	[5]
Coastal litter	Unmanned aerial vehicles and machine learning approach	To detect and map marine litter	[5]

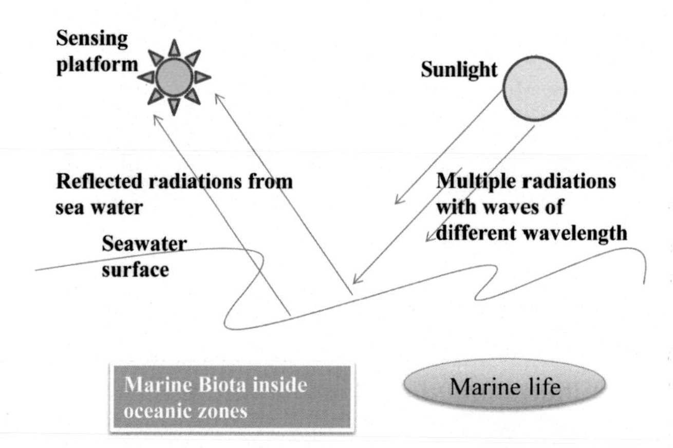

Figure 3.1 Remote sensing using radiation of different wavelength.

3.5 Analytical techniques in identifying pollutants

3.5.1 Basics in analytical techniques

Analytical techniques identify pollutants in marine water through some equipment in laboratories. This kind of technique is used for identifying pollutants that are dispersed or dissolved in the ocean. One such type of pollutant is microplastics. Monitoring of microplastics in abiotic and biotic environmental conditions can provide basic scientific information to determine their pollution status, concentration hotspots, historical trends and the exposure and fate of organisms. Analytical procedures for such pollutants in environmental samples require the following steps [6]:

- extraction;
- isolation;
- identification; and
- quantification.

Analysis of microplastics involves two steps. They are [6]:

- physical characterization of plastics, for example, microscopy; and
- chemical characterization, for example, spectroscopy

3.5.2 Physical characterization of marine pollutants identification

3.5.2.1 Microscopy

Previously stereomicroscopy was widely used for the identification of microplastics whose size fall in the hundreds of microns range. Synthetic and natural fibres are difficult to distinguish by microscope. Later scanning electron microscopy (SEM) was developed, which can provide extremely clear and high magnification images. These high-resolution images help to discriminate plastics from organic material. Energy dispersive X-ray spectroscopy (EDS) provides the elemental composition of samples. This helps to identify carbon-dominating plastics from inorganic ones [6].

3.5.3 Chemical characterization of marine pollutants identification

3.5.3.1 Fourier-transform infrared spectroscopy

Fourier-transform infrared spectroscopy (FTIR) provides details about the chemical bonds that are present in the particle that is isolated from water. Transmission, reflectance and attenuated total reflectance (ATR) modes are available for analysis. The ATR mode provides stable spectra for irregular microplastics. Nowadays, μ-ATR-FTIR is used for identification in environmental samples. The origin, source and input pathway of samples can be identified using FTIR [6].

3.5.3.2 Raman spectroscopy

Raman spectroscopy (RS) works using a laser beam which upon falling on a particle is reflected at different frequencies depending on the molecular structure and atoms present. It provides spectra similar to FTIR. It is sensitive to additives and pigments in microplastics that interfere with the identification of polymer types [6].

3.5.3.3 Carbon:Hydrogen:Nitrogen (C:H:N) analysis

Different groups of polymers possess distinctive elemental compositions, which are used to identify the plastic origin of a particle. For this purpose C:H:N analysis is performed. By doing so it can be assessed as whether the particles are from a polymer family. This technique is not applicable for smaller particles [7].

3.5.3.4 Thermal analysis

This analysis suggests that polymers change with respect to their physical and chemical properties, which are dependent on temperature. Differential scanning calorimetry (DSC) is used for studying the thermal properties of polymeric material. DSC is combined with thermogravimetry (TGA) for analyzing the polymers. It was useful in analyzing polyethylene and polypropylene. It is very fast but limited to certain polymers only [6]. Fig. 3.2 shows different analytical instruments used in identification of marine pollution.

3.6 Electrochemical detection of marine pollutants

3.6.1 Basics in electrochemical detection of marine pollutants

Pollution in the marine environment is an immense and scientific challenge to all researchers. The target is to identify the nature of the contaminants, along with their sources, distribution, concentration,

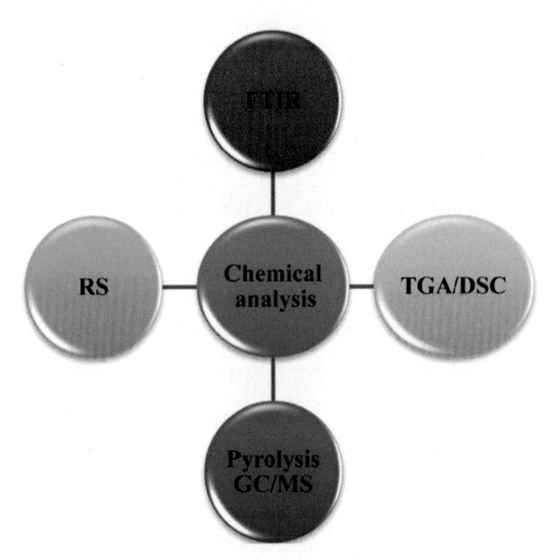

Figure 3.2 Analytical tools used in the identification of marine pollutants. *DSC*, Differential scanning calorimetry; *FTIR*, Fourier-transform infrared spectroscopy; *RS*, Raman spectroscopy; *TGA*, Thermogravimetry.

persistence, uptake into biota and effects on the ecosystem. In order to upgrade the identification, sensors are introduced. Sensors have been already identified for temperature, conductivity, depth and turbidity in ocean. Sensors are analytical devices incorporated with materials/biomaterials associated with or integrated with a physiochemical transducer that may be optical, electrochemical, thermometric, piezoelectric, or magnetic. There are different pathways used by pollutants in reaching marine water, for example, direct input (discharge from industries), riverine contributions (agricultural runoff) and drawdown from the atmosphere (emission from power plants) [8]. There are microbes that turn toxic in the marine environment due to effects of environment. Algal blooms and microbial contaminants can be toxic to the marine environment. It is found that harmful algal blooms adversely affect human health and the economy of nations. Since monitoring these harmful algal blooms by microscopy is very tedious, sensors have been developed. Sensors detect nucleic acids that are present in individual microbes and which vary between each family [9].

3.6.2 Structure of sensor

The sensor has three main components: receptor, transducers and final electrical signal. Fig. 3.3 shows the components in a sensor.

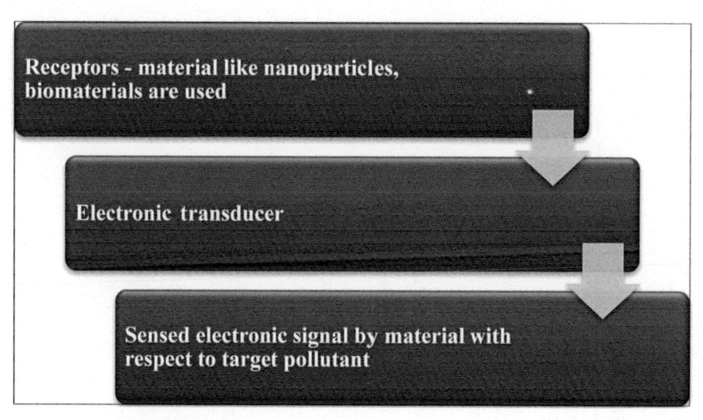

Figure 3.3 Structure of sensor.

3.6.3 Application of sensors for identifying marine pollutants

There are a few sensors that are derived from different materials like nanomaterials, polymer and biomaterials, which are applied in real-time applications like identifying pollutants in marine environments. Some of them are discussed in Table 3.3.

3.7 Computational models in detecting marine pollution

3.7.1 Introduction to computational models

There are numerous ways to detect marine pollutants. But it is important to identify certain mechanisms through which pollutants are killing the ecosystem, for example, from where pollutants are originating, where they are accumulating, what pathways they followed to reach marine life. To identify all these issues by sampling is not an easy task and it is time consuming. Pollutants may change morphological parameters by interacting with external agents like organisms. Hence new innovations are needed to identify pollutants as such. For that numerical models are introduced to estimate the sources, sinks and pathways of pollutants in the marine environment [13].

3.7.2 Objective of numerical modelling

There are two ultimate goals to improve the modelling of marine pollutants: (1) identifying where, how and why plastics enter and leave the

Table 3.3 Parameter identification using various sensors.

Parameter	Material	Immobilization	Methodology	References
Nitrate content	Enzyme nitrate reductase from various organisms	Polymer, sol–gel matrix	Optical transduction	[8]
Polycyclic aromatic hydrocarbon in fish tissue cells	Double standard calf thymus DNA	Buffer	Guanine oxidation peak	[10]
Heavy metals	Graphene/polyaniline/polystyrene	Electrospun nanofibres	Screen printed electrode, through anodic stripping voltammetry	[11]
Pesticides	Copper oxide attached with single-walled carbon nanotubes	Nanocomposites	Selective electrochemical detection	
Phenolic compounds	Chitosan modified gold electrode	Nanocomposites	Amperometric detection	
Phosphate	Cobalt phthalocyanine	Electrocatalyst	Amperometric detection	
Multiple compounds	Multiphysical chemical profiler	Gel integrated voltammetric microsensors and multichannel voltammetric probe	Microprocessor mechanism	[12]

ocean in different forms; (2) modelling the mass of desired pollutants and number of desired contaminants. Numerical modelling involves identifying interactions between various parameters, for example, fluxes that occur between the ocean and coast, movement between the coast to ocean interface and flux between the ocean and biota. There are two main components in modelling. One is the hydrodynamic part comprising circulation and wind wave and the other is oil slick transport [13].

3.7.3 Oil spill models

Numerous oil spill models have been developed to simulate the weathering process and forecast the fate of oil spilled. There are two types of oil spill models. One is oil weathering models to estimate how oil properties are changing with respect to time and the other type is deterministic models and three-dimensional models. Oil spill models are categorized into different generations. The first-generation model was used to trace the movement of rigid bodies under the influence of current and wind advection. Second-generation models were used to determine the forces and phenomena responsible for the fate and behaviour of chemicals in the sea. There are many models that are reviewed in many articles [14]. Some of them are listed below.

- General NOAA Operational Modelling Environment (GNOME)
- MEDSLIK-II
- SeaTrackWeb (STW)
- Model Oceanique de Transport d' Hydrocarbures (MOTHY)
- DieCAST-SSBOM (Shirshov-Stony Brook Oil spill transport Model)
- COastal Zone OIL spill model (COZOIL)
- Sea and Oil Slick 3D Simulation Model (SOSM)—a model used for transport and physiochemical evaluation of oil slicks
- North Aegean Sea and Oil Slick (NASOS) [14]

3.7.4 Plastics or marine debris model

Floating debris is modelled by tracking them from source to ocean based on realistic waste production and ocean currents. A model is being projected by researchers for analyzing this debris in ocean water. It is called hydrodynamic and particle dispersion modelling. It has two parts. One is studying the hydrodynamic nature of water using equations of motion and the other is analyzing the particle movement in flow using the hydrodynamic nature of surface water. The hydrodynamic nature of ocean water includes wind speed, heat flux and precipitation and is studied using a Hybrid Coordinate Ocean Model (HYCOM). The dispersion of floating material is studied using Pol3DD. The principle behind this model is the advection mechanism [15]. These models are helpful in identifying transport, floating and accumulation of debris and further formation of oceanic debris accumulation zones. Realistic input data are used in simulation models. They help to justify the impacts generated for more than decade. They advise the government to minimize waste generation and precautions are provided under extreme conditions. Hence simulation models are in high demand in economics.

3.8 Cyclodextrin-promoted fluorescence modulation

3.8.1 Promotion of fluorescence modulation

The ability to detect pesticides in the marine environment is highly needed because of their huge impact on the marine biota. Methods like mass spectroscopy, electrochemical assay and RS used in detection are highly time consuming, expensive, highly sensitive and require additional procedures prior to analysis [16].

3.8.2 Principle behind operation

Generally fluorescence detection involves the transfer of energy from an aromatic toxicant energy donor to a high-quantum-yield fluorophore acceptor, leading to a bright, turned on fluorescence signal. This is a method of using cyclodextrin to promote toxicant extraction followed by fluorescence detection. This can be used for the detection of both polar and nonpolar photophysically active toxicants. When the toxicant is not photophysically active and cannot transfer in energy transfer then fluorescence modulation can occur in which cyclodextrin promotes proximity-induced changes in the fluorophore emission signal when toxicants are close to the analyte. The detection of organic compounds in a real environment depends on the plethora of living species (both plant and animal) that can survive in high salinity (in saltwater systems), a broad range of pH, and can be found at a variety of temperatures [16].

3.8.3 General procedure for detection

Before detection, the general characterization of water samples is done using gas chromatography- mass spectroscopy (GC-MS). pH and conductivity is measured. It involves three stages for the analysis of pesticides in natural water, such as the marine environment [16].

3.8.3.1 Stage I

- Measurement of fluorescence modulation using the ratio of integrated emissions in the presence of analyte to the integrated emissions of fluorophore in the absence of analyte.
- Maintenance of temperature control in modulation.
- Calculating the limit of detection (LOD) using analyte concentration and fluorescence modulation ratio [16].

3.8.3.2 Stage II

Array analysis using SYSTAT 13 which is statistical computing software with the following settings.

- Classical discriminant analysis
- Grouping variable: analytes
- Predictors: α-cyclodextrin/4,4-difluoro-1,3,5,7,8-pentamethyl-4-bora-3a,4a-diaza-*s*-indacene (BODIPY), β-cyclodextrin/BODIPY, methyl-β-cyclodextrin/BODIPY, 2-hydroxypropyl-β-cyclodextrin/BODIPY, γ-cyclodextrin/BODIPY and PBS/BODIPY
- Long-range statistics: Mahal [16]

3.8.3.3 Stage III

Practical consideration involves selectivity for differentiating between structurally similar analytes, sensitivity for low concentrations of pesticides and general applicability for different water samples with varying salinity and pH, as well as for different temperatures of these water samples [16].

Cyclodextrin-promoted fluorescence modulation can be used for the detection of organochlorine pesticides in contaminated marine environments. The high sensitivity, selectivity and general applicability shows that this is a potable detection device that can be used in detecting contaminants of pesticides and oil spills in the marine ecosystem [16].

3.9 Algal biosensor—bioindicator for organic pollutants

Many microorganisms are used in the identification of organic pollutants in marine water through sampling, lab testing and analysis, which is time consuming, labour intensive, demands ex situ collection at locations, sample preparation and the cost depends on complexity. Hence biosensor and bioassay technology has been developed for monitoring marine pollutants with greater efficiency. When it comes to marine pollutants biosensors have to be fully automated, resistant to biofouling, highly resistant to corrosion and physical impacts, require minimal power consumption, be sensitive to measure pollutants at lower concentration and allow accurate calibration [17]. Several biosensors have been developed for measuring marine polluting components, such as eutrophication, antibiofouling agents, algal toxins, trace metals and organic compounds that are to be discussed in the upcoming section.

3.9.1 Principle behind operation—an algal biosensor

An algal biosensor mainly works on the measurement of photosynthetic activity. This is done by a photosynthetic organism and is based on the inhibition of the electron transfer occurring after a few minutes exposure of photosystem II (PSII) to certain pollutants, or to adverse physicochemical conditions changing the local chemical equilibrium. When pollutants such as photosynthetic pesticides are present and encounter the photosystem, they can bind the reaction centre D1 protein and directly or indirectly inhibit the transport of electrons from the primary acceptor, plastoquinone A (Q_A), to the secondary quinine (Q_B) along the photosynthetic chain. This inhibition results in a variation of PSII fluorescence emission in a pollutant concentration-dependent manner that can be monitored by optical transduction. But at certain conditions high salinity in the marine environment affects the photosynthetic process that shows insignificant changes in bioassay performance [17].

3.9.2 Stages in bioassay detection

- Growth of microalgae that are isolated from the marine environment.
- Fluorescence analysis—this measurement is composed of a fluorimeter that is provided with two red and two white light-emitting diodes (LEDs) and an optical fluorescence detector.

- Pesticide measurement—it is performed once the algae is exposed to different pesticide concentrations in seawater using standard instrument settings and calibration methods.
- LOD—it is determined based on the 99% confidence interval, which, assuming the normal distribution, corresponds to 2.6 × standard error of the measurements (σ).

$$LOD = 2.6 \times \sigma \times IC_{20}/(100 - 2.6 \times \sigma)$$

where IC_{20} is the 20% inhibitory concentration [17].

3.9.3 Use of algal bioindicator

Microalgae with lipid content has been suggested to have potential to accumulate in a high salinity environment which is mainly due to the presence of unsaturated lipids. Hence the presence of a lipid layer helps in the marine environment. Microalgae groups like Chlorophycea, Trebouxiophycea, Dinoflagellates, Diatoms and Eustigmatophycea groups were used in biosensor applications. They detect better than standard methods like GC-MS, as they provide precise measurement of the chosen pollutant with high accuracy and precision. They have great advantages like easy, low-cost and fast prescreening test of seawater samples, while providing real-time information about marine ecosystems [17].

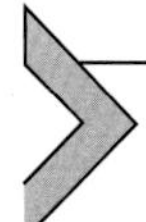

3.10 Bioluminescent bacteria—biomarker for organic pollutants

Bioluminescent bacteria are organisms that emit photons of λ_{max} 490 nm through a complex biochemical reaction in which the oxidation of long-chain aldehyde (C8−C14) and flavin mononucleotide ($FMNH_2$) takes place in the presence of a monooxygenase class of enzyme luciferase. They are present in the diverse environment ranging from marine ecosystems to terrestrial habitats. The blue−green photonic emission is due to the expression of *lux* genes which are commonly found in bioluminescent bacteria. There are five *lux* genes, that is, *luxA,B,C,D,E*. They are used for rapid identification of hazardous compounds like heavy metals and organic pollutants. It is noted that the toxicity of hazardous compounds may be due to [18]:

- high affinity toward carboxyl-, histidyl- and thioyl groups of cellular proteins and thereby altering their functional aspects;
- stimulated generation of reactive oxygen species (ROS); and/or
- disruption of vital transport channels through binding to specific sites [18].

3.10.1 Principle of operation of bioluminescent bacteria

The main working principle in bioluminescent bacteria is the *lux cassette*. *lux cassette* of luminescent bacteria encodes the components that give optical signals in the form of photons. When exposed to any stress caused by hazardous compounds, these bacteria tend to inhibit the optical signals rapidly. Therefore bioluminescent bacteria have widely been used for monitoring environmental pollutants, such as the synthetic antibiotic sulfamethoxazole and water through an optical-based sensing system with high sensitivity [18].

The bioluminescence method is quick and inexpensive and allows detection of a wide range of substances for subsequent accurate chemical analysis. The other advantages of bioluminescence tests are their high sensitivity to toxic substances, the simple measurement procedure, the availability of reagents and bioluminometers and the possibility of automating the analysis due to the use of immobilized reagents [18].

3.10.2 Development of bioluminescent bacteria for detection of heavy metals and pesticides

Researchers have developed bioluminescent bacteria immobilized on a suitable matrix for the detection of heavy metals and pesticides in waterways. The organism that is used for the enhanced production of luciferase enzyme is *Photobacterium leiognathi*. The organism is isolated from marine samples and grown in media and saved as inocula samples. Production of enzyme from this organism is optimized using various parameters. For safety purposes these organisms are immobilized in sodium alginate, agar and are fabricated as beads. A single bead was subjected to toxic analyte and luminescence response as relative light units (RLU) was measured with respect to time using a luminometer. The ratio of the lost/remaining RLU at any time to the initial RLU was referred to as relative light emission. The rate constant value, k_d, is calculated, from which the initial concentration of toxic pollutants can be found using graphical interpretation [18].

3.10.3 Uses of bioluminescent bacteria

Chemicals that are categorized as toxic agents with different properties adversely affect the cellular components or biochemical pathways. The effect of the bioluminescent system is the integrated effects of a number of cellular processes. Immobilized beads of *P. leiognathi* were found to retain the property of luminescence emission for more than 30 days under suitable conditions [18].

3.11 Microfluidic device integrated with algal fluorescence for pesticide detection

A microfluidic device can be defined as a combination of a set of microchambers and microchannels etched or moulded into different materials, such as glass, silicon and polymer. They are beneficial for pesticide detection, because they offer precise control of liquids and allow fluid manipulation under small volume and space. They also allow monitoring of rapid concentration changes of oxygen and carbon dioxide. This device is predominantly made of glass because glass is suitable for algae-based pesticide detection. Also it is nontoxic, biologically inert, optically transparent and impermeable for gases, therefore allowing the accurate determination of oxygen and other gases. The transparency of the glass enables illumination for photosynthesis. Also, algae can be removed from the glass microfluidic device and fresh algae can be refilled for reuse [19].

Although there are lots of methods for determining pesticides, this method is quite easy and less time consuming. The most compact idea for developing a fast, cheap and easy to handle detection platform was the use of cells (algae, bacteria, yeast and fungi) as biocatalytic elements. The interference and selectivity for pesticide detection with a cell-based platform are the drawback and advantages as well. To solve this issue algae are chosen for the detection of pesticides in water. Photosynthesis analysis and pesticide determinations based on the metabolism/photosynthesis of algae were studied by different groups with optical and electrochemical approaches. Electrochemical detection of pesticide concentration was provided with the support of electrochemical pH and oxygen sensors by analyzing algal photosynthesis and respiration. But the sensitivity through this detection is low. Hence optical detection was chosen in which a fluorescence-based oxygen sensor is the primary one. It was integrated with a miniaturized fluorescent sensor with a LED and photodetector into a microfluidic chip. Since the detection was low, a glass-based microfluidic device with integrated optical pH, oxygen and algal fluorescence detection for complementary analysis of pesticide concentration in a fast, cheap, reusable, and reliable manner has been introduced [19].

3.11.1 Steps in the fabrication of device

Fabrication of glass-based microfluidic device using a packed bed reactor design:

- Preparation of pH and oxygen sensor formulation using the aza-BODIPY dye (0.5 mg) and dry D4 hydrogel (50 mg). They were dissolved in tetrahydrofuran (THF) and polystyrene; silicone and PtTPTBPF dissolved in chloroform and toluene, respectively.
- The microfluidic device was silanized for better adhesion between the hydrogel matrix of the pH sensor and the glass surface.
- Luminescence lifetime measurements and dual luminescence referencing (DLR) were performed for oxygen and pH sensors, respectively.
- Algae solution was injected into the microfluidic device to form a biofilm at the solid–liquid interface.
- Integration of holder, microfluidics and detection system to develop the chip.
- Calibration of device and measurement of target pollutant in the samples [19].

3.11.2 Uses of microfluidic device with algal film

It is possible to analyze pesticide concentration in water in 2 min. It is possible to collect complementary information about algal photosynthesis behaviour and pesticide concentration with pH, oxygen sensors and algal fluorescence in buffered solutions. The device can be fabricated for use based on the saline nature in marine ecosystems and can be optimized to reduce the total analysis time [19].

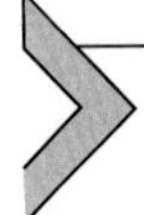

3.12 Application of ultraviolet fluorescence spectroscopy—oil/mineral aggregate formation

For the past 26 years ultraviolet fluorescence spectroscopy (UVFS) has been used to estimate crude oil aggregate in natural water. UVFS is a sensitive detector for oil pollutant in water. This is done after performing solvent-extraction. This process helps in different extraction processes, thereby concentrating natural oil that can be analyzed using fluorescence spectroscopy [20]. UVFS is generally applied for the removal of oil from oil–water–sediment aggregate. As a result, if UVFS is to be applied as a tool to predict and monitor the multiphase system produced by sediment oil aggregation in the waters of a particular clean-up site, it has to be applied as direct fluorescence measurements on unextracted samples [20].

3.12.1 Sample application process

UVF spectra of both aggregates and dissolved/dispersed oil were obtained using the spectrafluorometer with excitation wavelength set at 320 ± 2 nm. The suspension was immediately scanned between 340 and 500 nm to ensure that artefacts associated with the settling of aggregates did not occur. The same protocol was followed to determine the fluorescence from 340 to 500 nm in a dissolved/dispersed fraction with care taken to prevent the resuspension of aggregates [20].

3.12.2 Basic code behind in detecting oil–mineral aggregates

Detection of oil–mineral aggregates is associated with distinct shifts in UVFS emission spectra. The emission fluorescence spectra for dispersed oil is between 330 and 430 nm, whereas for oil/mineral aggregate there is a broad fluorescence peak at 350 nm. The remaining supernatant seawater that contained dissolved/dispersed components of the oil had a smaller, narrower peak of fluorescence at 355 nm. Most unsubstituted aromatic hydrocarbons exhibit intense fluorescence in the ultraviolet or visible region of the electromagnetic spectrum. As the degree of conjugation of the aromatic groups in oil increases, the intensity of fluorescence increases, thereby inducing a shift in fluorescence. Additionally changes in chemical composition of oil ranging from lower molecular weight to higher molecular weight may increase the oil viscosity, thereby reducing the intensity in oil/mineral aggregate spectra [20].

3.12.3 Interference in ultraviolet fluorescence spectroscopy

Quantitative direct microscopic measurements were used to determine the fluorescent area of oil/mineral aggregate, which confirmed that low viscous oil formed a good number of oil/mineral aggregates, thereby producing an exponential increase in fluorescence emission at 450 nm. Hence this inverse relationship between oil/mineral aggregate and viscosity plays a crucial role in emission spectrum analysis [20].

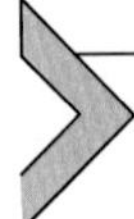

3.13 Direct observation of marine debris—application of various platforms

In order to characterize the abundance of different types of marine debris various observing platforms are needed. Each platform has its own advantages and disadvantages in terms of level of temporal and spatial resolution, the sensors and samplers they can carry, debris types and the size they can cover. Some of the platforms are discussed below [21].

3.13.1 Satellites, aircraft and drones

Aerial survey and remote technologies provide rich information on debris at the ocean or shoreline surface. The big part of marine debris is submerged in the ocean and lakes or buried in the sand and rocks, but direct measurements remain critical for comprehensive monitoring. Hence sensors are used in field monitoring systems to gain in-depth knowledge of various geospatial and temporal scales [21]. Sensors have certain limitations that are related to spectral resolution, spectral range, sensitivity, revisit time, geospatial resolution and coverage. To avoid this integrated sensing applications have been introduced in order to involve sensors of different types [21]. Some of the sensors that are available to identify plastic debris in the ocean are high spatial resolution imaging, optical–spectroradiometric technique and radar sensors. Washed ashore, floating and slightly submerged marine debris has been monitored using high-resolution cameras on fixed platforms. This is used by applying a visible spectrum to make true colour RGB composites images. It can very specifically detect what kind of debris or specific object is floating in ocean. But RGB images do not provide the physical and chemical composition of the litter [21]. Spectroradiometric analysis from the ultraviolet to the far-infrared spectrum has opened new avenues for the detection and characterization of plastic and other types of marine debris. They provide very clear information about plastics and their features. Spectral information from multi- and hyperspectral optical sensors can be used to infer the abundance of plastic objects of a subpixel size. A prospective technique to observe submerged debris is active remote sensing using a light detection and ranging system (LIDAR) that can measure the onboard laser lights backscattered from the ocean [21].

3.13.2 Ships

Ships are traditionally used to collect data on marine debris floating on or near the surface. They provide platforms for a variety of sensors and samplers for a comprehensive study of the entire water column, from the seabed to the surface [21].

3.13.3 Autonomous platforms

Floats, gliders (both seagliders and wavegliders) and autonomous surface and underwater vehicles [autonomous surface vehicle (ASVs) and autonomous underwater vehicle (AUVs)] equipped with adequate sensors can provide unique measurements in hard-to-reach parts of the ocean [21].

3.13.4 Fixed point observations

These are important and efficient platforms to monitor temporal variability and, in particular, long-term trends of the problem. There are nearly 120 open ocean observation sites and a higher number of coastal and shelf observation that can be used for marine debris observation [21].

3.13.5 Benthic landers and crawlers

These provide unique information about the arrival of debris on the seabed and the interaction with this rather different benthic ecological community. They are devices that remains on the seabed for a protracted period of time [21].

3.13.6 Shoreline monitoring and beachcombing

The highest concentrations of marine debris are reported in shorelines that are cleaned by continuous clean-up using floaters [21].

3.14 Conclusion

The chapter briefly covers different tools used for identifying waste in the marine environment. Various tools like remote sensing, analytical equipment, sensors and simulation models are discussed. It can be understood that all detection tools should be interlinked with each other to measure the pollutants in marine life. Complete impact studies for a chosen area can be covered by remote sensing but it involves a spectrum response at the desired wavelength. Hence only visible objects can be tooled. Whereas in sensors and analytical equipment only target pollutants like microplastics and heavy metals can be deduced at the laboratory scale. Simulation models are used to assess the impact caused due to certain pollutants, such as floating debris in surface water. Each detection tool has its own advantages and disadvantages. It is advised to interconnect all tools for the identification of marine contaminants.

References

[1] Dahms HU. The grand challenges in marine pollution research. Front Mar Sci 2014;1. Available from: https://doi.org/10.3389/fmars.2014.00009.

[2] Wolfe D. Marine pollution monitoring: objectives and design criteria. OCEANS 1987;87. Available from: https://doi.org/10.1109/oceans.1987.1160573.

[3] Clark CD. Satellite remote sensing for marine pollution investigations. Mar Pollut Bull 1993;26(7):357–68. Available from: https://doi.org/10.1016/0025-326x(93)90182-j.

[4] Park PK, Elrod JA, Kester DR. Applications of satellite remote sensing to marine pollution studies. Chem Ecol 1991;5(1-2):57–73. Available from: https://doi.org/10.1080/02757549108035244.

[5] Hafeez S, Sing Wong M, Abbas S, Kwok CYT, Nichol J, et al. Detection and monitoring of marine pollution using remote sensing technologies. Monit Mar Pollut 2019. Available from: https://doi.org/10.5772/intechopen.81657.

[6] Shim WJ, Hong SH, Eo SE. Identification methods in microplastic analysis: a review. Anal Methods 2017;9(9):1384–91. Available from: https://doi.org/10.1039/c6ay02558g.

[7] Löder MGJ, Gerdts G. Methodology used for the detection and identification of microplastics—a critical appraisal. Mar Anthropogenic Litter 2015;201–27. Available from: https://doi.org/10.1007/978-3-319-16510-3_8.

[8] Kröger S, Piletsky S, Turner APF. Biosensors for marine pollution research, monitoring and control. Mar Pollut Bull 2002;45(1-12):24–34. Available from: https://doi.org/10.1016/s0025-326x(01)00309-5.

[9] LaGier MJ, Fell JW, Goodwin KD. Electrochemical detection of harmful algae and other microbial contaminants in coastal waters using hand-held biosensors. Mar Pollut Bull 2007;54(6):757–70. Available from: https://doi.org/10.1016/j.marpolbul.2006.12.017.

[10] Bagni G, Baussant T, Jonsson G, Barsiene J, Mascini M. Electrochemical device for the rapid detection of genotoxic compounds in fish bile samples. Anal Lett 2005;38 (15):2639–52. Available from: https://doi.org/10.1080/00032710500371105.

[11] Hernandez-Vargas G, Sosa-Hernández J, Saldarriaga-Hernandez S, Villalba-Rodríguez A, Parra-Saldivar R, Iqbal H. Electrochemical biosensors: a solution to pollution detection with reference to environmental contaminants. Biosensors 2018;8 (2):29. Available from: https://doi.org/10.3390/bios8020029.

[12] Zielinski O, Busch JA, Cembella AD, Daly KL, Engelbrektsson J, et al. Detecting Marine Hazardous Substances and Organisms: Sensors for Pollutants, Toxins, and Pathogens. Marine Science Faculty Publications; 2009. p. 159. Available from: https://scholarcommons.usf.edu/msc_facpub/159.

[13] Hardesty BD, Harari J, Isobe A, Lebreton L, Maximenko N, Potemra J, et al. Using numerical model simulations to improve the understanding of micro-plastic distribution and pathways in the marine environment. Front Mar Sci 2017;4. Available from: https://doi.org/10.3389/fmars.2017.00030.
[14] Zafirakou A. Oil spill dispersion forecasting models. Monit Mar Pollut 2019. Available from: https://doi.org/10.5772/intechopen.81764.
[15] Lebreton LC-M, Greer SD, Borrero JC. Numerical modelling of floating debris in the world's oceans. Mar Pollut Bull 2012;64(3):653–61. Available from: https://doi.org/10.1016/j.marpolbul.2011.10.027.
[16] DiScenza DJ, Lynch J, Miller J, Verderame M, Levine M. Detection of organochlorine pesticides in contaminated marine environments via cyclodextrin-promoted fluorescence modulation. ACS Omega 2017;2(12):8591–9. Available from: https://doi.org/10.1021/acsomega.7b00991.
[17] Moro L, Pezzotti G, Turemis M, Sanchís J, Farré M, Denaro R, et al. Fast pesticide pre-screening in marine environment using a green microalgae-based optical bioassay. Mar Pollut Bull 2018;129(1):212–21. Available from: https://doi.org/10.1016/j.marpolbul.2018.02.036.
[18] Ranjan R, Rastogi NK, Thakur MS. Development of immobilized biophotonic beads consisting of Photobacterium leiognathi for the detection of heavy metals and pesticide. J Hazard Mater 2012;225-226:114–23. Available from: https://doi.org/10.1016/j.jhazmat.2012.04.076.
[19] Tahirbegi IB, Ehgartner J, Sulzer P, Zieger S, Kasjanow A, Paradiso M, et al. Fast pesticide detection inside microfluidic device with integrated optical pH, oxygen sensors and algal fluorescence. Biosens Bioelectron 2017;88:188–95. Available from: https://doi.org/10.1016/j.bios.2016.08.014.
[20] Kepkay PE, Bugden JBC, Lee K, Stoffyn-Egli P. Application of ultraviolet fluorescence spectroscopy to monitor oil–mineral aggregate formation. Spill Sci Technol Bull 2002;8(1):101–8. Available from: https://doi.org/10.1016/s1353-2561(02)00122-6.
[21] Maximenko N, Corradi P, Law KL, Van sebille E, Garaba SP, Lampitt RS, et al. Towards the integrated marine debris observing system. Front Mar Sci 2018;6:447. Available from: https://doi.org/10.3389/fmars.2019.00447.

CHAPTER FOUR

Oil spill clean-up

4.1 Introduction to oil spills and its contamination

Generally oil spills are discussed by lots of researchers as they have crucial ecological effects on marine life and the surrounding environment. Basically crude oil and petroleum products are a mixture of polycyclic aromatic compounds and hydrocarbons. They find their way into ocean through industrial effluents and petrochemical product leakage by ships, oil well drilling production operation, downstream industry and gas pipelines. Oil is a mixture of organic compounds enriched with hydrogen bonded with carbon as the backbone. This chemical chain is difficult to break. Oil spills in the environment may affect organisms by direct toxicity or by physical smothering. Oil in water may deplete dissolved oxygen due to the transformation of organic compounds to inorganic compounds that further leads to the loss of biodiversity. A few health effects, such as toxicity in fishes, lymphocytosis, epidermal hyperplasia and haemorrhagic septicaemia, can be caused by oil spills. Another toxicity which is witnessed due to oil spills is photoenhanced toxicity which leads to the activation of chemical residues that bioaccumulate in marine organisms. Furthermore, oil spills can contaminant soil by affecting parameters such as minerals, organic matter and pH. Oil creates anaerobic condition in the soil, coupled to water logging and acidic metabolites, which leads to the high accumulation of aluminium and manganese ions that are toxic to plant growth [1]. This chapter deals with some of the treatment methodologies that are followed for removing oil spill contamination from ocean water.

4.2 Treatment methodologies followed for removing oil spills from marine water

Some of the common technologies that are adopted for cleaning oil spills are cleaning scrubbing, vacuuming, low- and high-pressure washing,

Modern Treatment Strategies for Marine Pollution.
DOI: https://doi.org/10.1016/B978-0-12-822279-9.00009-9

bioremediation and using sorbents. Nowadays new innovations have been developed due to new findings by researchers such as the use of nanomaterials, removal using biosurfactants and fibres and many other technologies. The adaptation of technology relies on its simplicity, efficiency in removing oil spills from water and cost effectiveness [2]. Fig. 4.1 shows some of the common techniques that are used for cleaning oil spills.

4.3 Oil spill removal using sorbents

4.3.1 Introduction to sorption

Sorption is a popular technique for cleaning up oil spills. Adsorption is a simple, inexpensive tool for removing oil spills from marine water. There are a few characteristics to be followed before choosing an adsorbent for cleaning up oil spills, such as the adsorbents should be oleophilic and hydrophobic in nature in order to attract oil from water. Some of the characteristics are listed below [3].

- Rate of adsorption
- Adsorption capacity
- Oil retention
- Ease of application

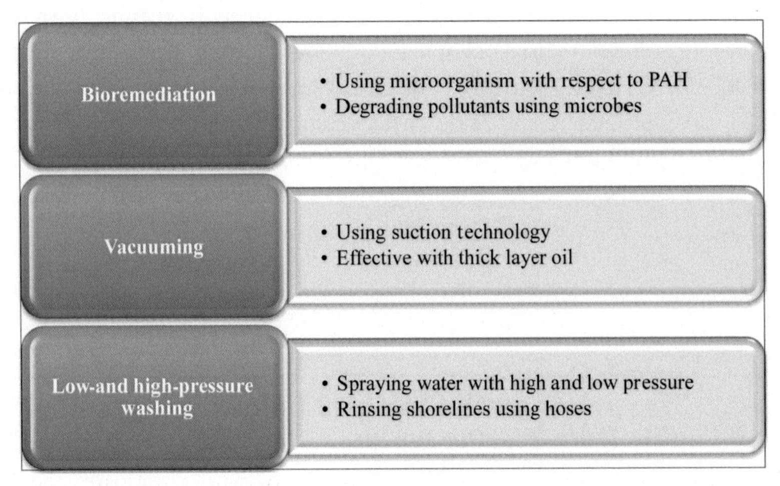

Figure 4.1 Common technology adopted in cleaning oil spills. *PAH*, Polycyclic aromatic hydrocarbon.

- Kinetic model and equation used for calculating isotherm [3]

4.3.2 Classification of adsorbents

Sorbents can be classified based on basic categories. They are:

- Organic and agro-based products
- Synthetic material
- Inorganic material

All these sorbents differs in recyclability, wettability, density, geometry and sorption capacity. The major problem associated with the use of sorbents is that it can be labor- and time-consuming. The major interfering factors that affect the sorbent are the increase in oil and emulsion density that reduces the buoyancy of thesorbents. Changes in emulsion visocity will interfere with sorbent effectiveness [4].

4.3.3 Organic and agro-based products

Natural or agro-based products are used as adsorbents, for example, straw, sawdust, peat, corn stalk and cotton fibres. Products that are derived from biomass, such as plants and animals, are termed organic adsorbents. Cellulose, hemicellulose and lignin are used as biowaste. These sorbents are nontoxic, noncorrosive, low-cost and active after recycling. There are many studies portraying the use of adsorbents for adsorbing oil. Oil uptake capacity, buoyancy and solubility of hydrocarbon in water are used as indicators to detect the sorption efficiencies. Agrowaste such as sugarcane leaves, straw and sugarcane bagasse are used for adosrbing oil from marine water. It is well-known that sugarcane bagasse has a high oil sorption capacity with respect to its particle size and contact time [3]. These adsorbents are environment-friendly and biodegradable. But they are hydrophilic in nature and this results in a loss of sorbent buoyancy [3].

4.3.4 Synthetic adsorbents

Synthetic adsorbents are prepared from agricultural products and wastes, household wastes, industrial wastes, sewage sludge and polymeric adsorbents. Some of the synthetic materials, such as polypropylene and polyurethanes, are the most commonly used commercial sorbents due to their oleophilic and hydrophobic properties. They have their own properties like porosity and pore structure. They are not biodegradable and their disposal through landfill is not advisable, whereas incineration is costly and banned in many places. They are not naturally occuring materials [3].

4.3.5 Inorganic adsorbents

The need for improved and high adsorption capacity at low cost led to a dependence on inorganic adsorbents like charcoal, clay minerals and zeolites. Among these zeolites gained wide application in environmental remediation. Mesoporous silica materials with surfactants showed better removal of organic pollutants from water. Some of these materials are hydrophilic in nature, thus they show high adsorption capacity [3].

4.3.6 Nanoparticles as adsorbents

Nanoparticles include carbon nanotubes (CNT), graphene and nanometal oxides that are commonly using to eradicate organic pollutants from water due to their chemical bonding ability. CNTs are highly hydrophobic in nature, hence they have a high adsorption capacity to enhance the deoiling process. Reseachers have developed CNT sponges which can adsorb oil and be reused. Carbon aerogels, graphene foams and porous carbon nanoparticles have been developed that showed higher adsorption capacity [3]. An enhanced adsorption capacity was noted for a graphene multilayer 3D structure that was used to remove oil spills from water. An ideal adsorbent to be used for removing oil from water show possesses the following characteristics [5]:

- Hydrophobicity
- High absorption and retention capacities
- Good selectivity
- High oil absorption rate
- Good mechanical resistance
- Reusability
- Buoyancy
- Low cost
- Abundance and biodegradability

4.3.7 Biosorbents

Biologically derived compounds like cellulose and chitosan are useful for removing oil-based pollutants from water. These are called biopolymer-based sorbents. Among them chitosan has proven to bethe most efficient biopolymers for the removal of oil spills from water. It is biodegradable, biocompatible, eco-friendly and low cost and has a high sorption capacity towards oil, as well as a unique structure allowing chemical functionalization. Chitosan is available in various forms, such as chitosan flakes, chitin flakes, chitosan powder and naturally derived chitosan from shrimp shells. Chitosan composites with metals and metal oxides also show good adsorption capacity compared to other sorbents. Chitosan can be combined with microorganism to produce microcosm inoculated chitosan, which showed 60% removal of pollutant from water [6]. A study showed that fish scales by themselves can be used as biosorbents for the removal of oil spills from water and showed a maximum removal efficiency of 93% [7].

4.4 Microbial degradation of petroleum hydrocarbons

4.4.1 Introduction to polycyclic aromatic hydrocarbons and its effects in marine biota

Pollution due to hazardous materials and its adverse effects on human health has been gaining more attention in the research field to eradicate pollutants completely and save the Earth. Among them polycyclic aromatic hydrocarbons (PAHs) have top place in the ranking of organic pollutants because of their toxic, mutagenic and carcinogenic properties. PAH are aromatic rings with linear, angular, or cluster arrangements. Electrochemical stability, persistency, resistance to biodegradation and a carcinogenic index of PAHs increase with an increase in the number of aromatic rings, structural angularity and hydrophobicity, while volatility tends to decrease with increasing molecular weight. They have a tendency to be bioaccumulated in food chain. Based on their toxicity, nearly 16 PAHs are listed as environmental pollutants by the US Environmental Protection Agency (US-EPA). There are two types of PAH: low-molecular-weight PAH (containing two or three aromatic rings) and high-molecular-weight PAH (containing more than four aromatic rings) [8].

PAH can covalently bond with DNA, RNA and proteins that are present in the biota. Low-molecular-weight PAH is toxic, whereas high-molecular-weight PAH is genotoxic. There are lots of technologies available for cleaning up the polluted places like UV oxidation and solvent extraction. Though the above techniques are worth in destroying contaminated compounds, they transfer the pollutant from one environment to other. It is an efficient and eco-friendly cleanup that breaks down the toxic compounds into harmless products. There are two types of treatment, namely, aerobic and anaerobic treatment. Among them anaerobic treatment studies have been reported frequently due to their merits [8].

4.4.2 Factors influencing bioremdiation of polycyclic aromatic hydrocarbons

There are numerous biotic and abiotic factors that affect the removal of PAH by microorganisms. Those are microorganisms' growth with respect to the surrounding environment, PAH and its nature and other physical parameters. These factors differ from site to site, which in turn can influence the process of bioremediation either by inhibiting or improving the growth of pollutant-degrading microorganism in the chosen environment [8].

4.4.2.1 *Temperature*

When temperature increase the solubility of PAH increases which in turn increases its bioavailability of PAH molecules. This in turn affects dissolved oxygen content in water, thereby disturbing the metabolic activity of aerobic mesophilic microorganisms. Hence with respect to temperature microorganisms are chosen like thermophiles which can work at high temperature, whereas psychrophiles can operate at low temperature. Examples are *Thermus* and *Bacillus* spp [8].

4.4.2.2 *pH*

pH plays a major role in the biodegradation of organic pollutants like PAH. Generally microorganisms are pH sensitive. Microbes attain their normal activity at pH 6.5–7.5 [8].

4.4.2.3 *Oxygen*

The biodegradation of organic pollutants like PAH can work under aerobic and anaerobic conditions. During aerobic conditions oxygen is needed for breaking aromatic chains in the reaction. But in anaerobic conditions oxygen is not required for degrading PAH, it depends on enzymatic activity [8].

Other factors like nutrients and bioavailability of PAH in the environment decide the degradation of PAH in environment. Nutrients like carbon, nitrogen and phosphorous are needed for the growth of microbes in the environment. The bioavailabilty of nutrients or PAH depends on dissolution, desorption, diffusion and hydrological processes like mixing and metabolism [8]. Fig. 4.2 shows common factors affecting bioremediation.

4.4.3 Aerobic degradation of polycyclic aromatic hydrocarbon

During aerobic degradation of PAH by bacteria, PAH is oxidized to diols through the addition of oxygen in to PAH. These diols are further oxidized to catechols and PAH rings are cleaved by dioxygenase enzyme. PAH (unsubstituted) with high thermodynamic stability can be broken into compounds that are less stable than the parent compunds. During aerobic degradation the addition of humic sustances will enhance the degradation of PAH. Also the addition of certain stimulants like salicylic acids has increased the production of the enzyme dioxygenase [9].

4.4.4 Anaerobic degradation of polycyclic aromatic hydrocarbon

When demand for oxygen exceeds the environmental supply the anaerobic tranformation of PAH plays a major role. Anaerobic degradation is mediated by denitrifying or sulfate-reducing bacteria [9].

4.4.5 Common microbes involved in polycyclic aromatic hydrocarbon degradation

Fungi and bacteria work in combination for the degradation of PAH in the environment. Some fungi can use cytochrome P450 to transform PAH and activate PAH for use by bacteria for further degradation [10] . Many microbes are involved in PAH degradation. There is fungal degradation which involves the tranformation of PAH by monooxygenase. There are two variaties of fungi: ligninolytic fungi and nonligninolytic fungi. Compared to bacteria and fungi, degradation by microalage (cyanobacteria and diatoms) has been paid less attention. It is found that microalgae are dependent on light intensity which purely determines the

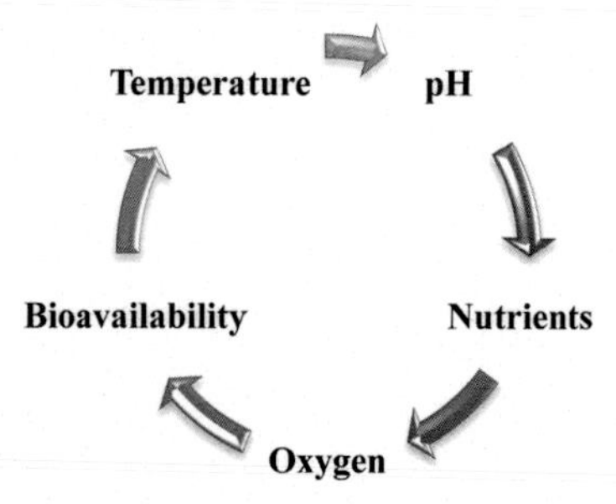

Figure 4.2 Factors affecting bioremediation.

degradation capacity of PAH. Many studies demonstrated the use of gold and light intensity as side parameters. Various white-rot fungi can metabolize PAHs along with bacteria in consortia by improving the bioavailability of target compounds. Due to a lack of suitable enzymes, fungi cannot degrade high molecular weight hydrocarbons (HMW) PAHs fully, but can transform them into polar metabolites with their extracellular enzymes, which can further be degraded by bacteria and other microbes [8]. Degradation using various microbes with respect to catgeories of PAH is discussed in Table 4.1

4.4.6 Fertilizers enhanced biodegradation

As seawater is a poor source of nutrients like nitrogen and phosphorous, bioremediation empolying fertilizers to increase the concentration of these nutrients is needed for the growth of hydrocarbon-degrading microorganisms. Fertilizer-enhanced biodegradation can be used for physical cleanup and surface and subsurface porous sediments. The application was empolyed in the Exxon Valdez spill where about 48,400 kg and 5200 kg of nitrogen and phosphorous, respectively, was applied for bioremediation. The results showed that microbes degraded 90% of alkanes and 36% of the initial mass of oil in 20–36 days. This showed about a threefold enhancement of biodegradation rate compared to unfertilized controls [4].

Field tests were conducted at an oiled shoreline in Price Willian Sound. It was examined with three types of fertilizers: (1) a water-soluble fertilizer, typical of what would be used in a garden; (2) a solid, slow-release fertilizer that would gradually release nutrients; and (3) an

Table 4.1 Category of polycyclic aromatic hydrocarbon (PAH) and microbes.

Category of PAH	Microbes	Nature of microbes	References
Low- and high-molecular weight PAH	*Mycobacterium* sp.	Bacteria uses PAH as source of carbon and energy	[10]
Low- and high-molecular weight PAH	*Bjerkandera* sp.	White-rot fungi	[10]
Low- and high-molecular weight PAH	*Penicillium janthinelum* and bacterial consortium	Nonligninolytic fungus and bacterial consortium	[10]
High-molecular weight PAH	*Selenastrum capricornutum* and *Scenedesmus acutus*	Microalgal species	[8]

oleophilic fertilizers to adhere to the oil surface. These three fertilizers were chosen based on application strategies, logistical issues for large-scale application, commercial availability and the ability to deliver nitrogen and phosphorus to surface and subsurface microbial communities for sustained periods. The results showed that treated areas looked clean in air but didn't meet scientific standards, although it gained public and political support. Additionally it showed that the rate of oil degradation was critically dependent on the ratio of nitrogen to biodegradable oil and oxygen is one limiting factor. Biodegradation rates for PAHs could increase by a factor of two, and for aliphatic hydrocarbons by a factor of five, with fertilizer [4].

4.4.7 Efficacy in the use of bioremediation

Bioremediation and natural oil biodegradation are not effective in all environments. Bioremediation was shown to be effective in highly porous shorelines where nutrients and oxygenated seawater could reach the surface and subsurface oil residue. Also it will not result in the complete removal of oil spills. The decision to use bioremediation should be based on net environmental benefit analysis. If the floating oil does not possess any ecological isssue then it can be left as such to undergo natural bioremediation. Scaling up is a critical factor which depends on logistical consideration and monitoring to ensure its effectiveness and toxicity development. The site and surface morphology has be to considered before applying microbes for cleaning [4].

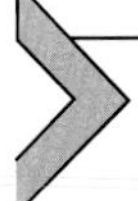

4.5 Application of biosurfactants in removing polycyclic aromatic hydrocarbon

4.5.1 Introduction to biosurfactants

Bacteria produce biosurfactants to facilitate microbial oil spill uptake and facilitate degradation by emulsifying hydrocarbons. Biosurfactants-producing microorganisms create their own environment and induce emulsification through various mechanisms. Biosurfactants-producing microorganisms are biologically active because of their versatile properties, minute toxicity and biological acceptability. Biosurfactants induce degradation of contaminants by increasing the surface area of substrates.

Biosurfactants emulsify the compounds, increase the water solubility and make the pollutant more accessible for the microorganisms. The use of surfactants may increase the hydrocarbon bioavailability which promotes the rate of biodegradation. Biosurfactants have higher surface activity with high tolerance and can withstand various environmental factors from mean to extreme conditions such as acidity or basicity of an aqueous solution, temperature, salt concentration, ionic strength, biodegradable nature, demulsifying–emulsifying ability, antiinflammatory potential and antimicrobial activity. Biosurfactants are amphiphilic compounds consisting of hydrophilic polar moieties, such as monosaccharides, proteins, polysaccharides, or peptides and the hydrophobic moiety carries either unsaturated or saturated fatty alcohols or hydroxylated fatty acids. One of the important features in biosurfactants is the hydrophilic–lipophilic balance [11].

Surfactants behave as excellent foaming agents, emulsifiers and dispersing agents with respect to their surface area. Due to their amphiphilic structure, biosurfactants increase the ability to change the property of the cell surface of microorganisms, along with the hydrophobic substance's surface area. They show selectivity to degrade the substrate and are functionally active in extreme conditions. There are many properties that make them effective in bioremediation and other technologies, such as dispersion, wetting, emulsification or deemulsification and foaming. Different microorganisms produce various types of surfactants that are classified below [11].

4.5.2 Classification of biosurfactants

Biosurfactants are classified based on its chemical composition and on the basis of the origin of microbes [11].

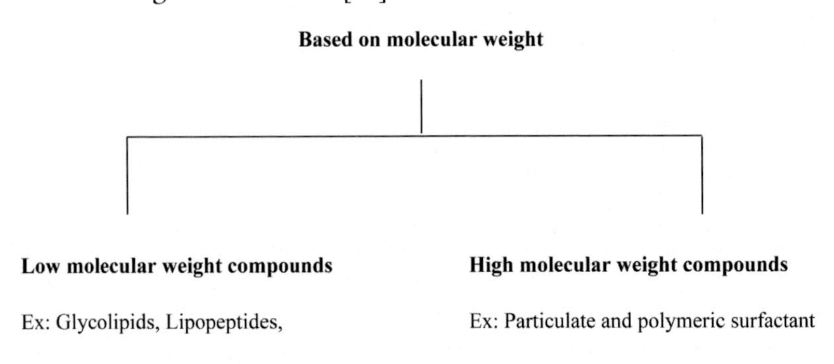

Table 4.2 gives sources of surfactants from various microbes.

4.5.3 Properties of surfactants

There are some properties that define the activity of biosurfactants in degrading toxic compounds like PAH. They are:

- Surface and interface activity
- Temperature
- pH
- Ionic strength
- Biodegradability
- Biofilm formation

4.5.4 Oil remediation using biosurfactants

Remediating oil spills using surfactants is effective because they try to break the compound and dissolve them in water. Much of the literature has suggested that biosurfactants are capable of destroying oil slicks which float on the surface of water and promote the dispersion of oil in water by forming a stable emulsion which enhances the rate of biodegradation. All these factors show the potential of biosurfactants for cleaning oil spills from water [11]. Biosurfactants produced by bacteria have a greater tendency to degrade hydrocarbons as they utilize them as their nutrients. The marine bacteria which degrade hydrocarbons are called hydrocarbonoclastic bacteria, as they have greater tendency to degrade both the aliphatic and aromatic fraction of crude oil. There are many factors that decide the degradation capacities of surfactants like the aerobic or anerobic environment, temperature, pH, nutrients, inorganic ions and bioavailability [11].

4.6 Sponges in removal of oil spills

Due to the economic crisis and the health hazards aspects of the release of oil spills, the removal of these pollutants by novel technology is needed. This issue has stimulated the synthesis of new materials for water—oil separation. This led to the development of material like sponges for large-scale removal of oil from marine water systems. The main property that helps the removal of oil from water by sponges is superhydrophobicity and superoleophilicity. Meshes, films and membranes have been fabricated in order to address this problem. These products were not applied for mass applicability as the removal was quite difficult. Hence sponges

Table 4.2 Sources and types of surfactants.

Surfactants	Examples	Microorganism producing them	References
Glycolipids	Sophrolipds Rhamnolipids Trehalolipids	*Corynebacterium* sp., *Mycobacterium* sp. and *Nocardia* sp, *Torulopsis bombicola, Bacillus licheniformis, Acinetobacter* sp, *Rhodococcus erythropolis*	[11,12]
Lipopeptides	It contain lipid as its functional group	*T. bombicola, B. licheniformis, Bacillus subtilis*	
Phospholipids	Phospholipid structure	*T. bombicola, B. licheniformis, Thiobacillus thiooxidans*	
Lipopetide and lipoprotein	Peptide-lipid	*B. licheniformis*	[12]
	Viscosin	*Pseudomonas fluorescens*	
	Serrawettin	*Serratia marcenscens*	
	Surfactin	*B. subtilis*	
	Subtilisin	*B. subtilis*	
	Gramicidin	*Bacillus brevis*	
	Polymyxin	*Bacillus polymyxia*	
Polymeric surfactant	Emulsan	*Acinetobacter calcoaceticus*	[12]
	Biodispersan	*A. calcoaceticus*	
	Liposan	*Candida lipolytica*	
	Carbohydrate–lipid–protein	*P. fluorescens*	
	Mannan–lipid–protein	*Candida tropicalis*	
Particulate surfactant	Vesicles	*A. calcoaceticus*	[12]

have been evolved. A study by Calcagnile et al. [13] introduced new material as sponges with magnetic property. Polyurethane foams functionalized colloidal superparamagnetic iron oxide nanoparticles and polytetrafluoroethylene particles to separate oil from water emulsion. As these foams are of light weight they float easily on water which can be removed from water using magnets. Such foams can be scaled up to clean spills in the marine environment. Generally polyurethane sponges show high absorption capacity, low cost and good elasticity. Along with oil, water gets absorbed which can be modified by altering the wetting property, for example, by coating these sponges with hydrophobic trimethylchlorosilane or tetraethoxysilane [14].

Apart from polyurethane there are certain nanofibres, such as CNTs and graphene, that have been fabricated in the form of sponges and tested for the removal of oil spills from water. Sponges work based on absorption which is simple, low-energy consumption, with high efficiency and uses organic materials that are very difficult to degrade. This technology is quite simple compared to chemical or biological works. Biological degradation consumes time whereas chemical degradation is very tedious [15].

4.7 Floating foams in cleaning up oil spills

There are lot of oil sorbents that have been used for remediating oil spills from water, such as polystyrene fibre and polypropylene fibres. These are called polymeric foams as they are composed of a polymer matrix and a gaseous phase. They have been applied in many environmental remediating applications and their efficiency lies in the development of superhydrophobic and superoleophilic polymeric foams [16]. However, they are costlier, with low oil uptake capacities, time-consuming and complicated fabrication methods. Hence an efficient, cheaper and easily scalable solution has been developed that is floating or modified floating foams. The most promising foams are polyurethane and melamine that are commonly used in commercial works. These are modified to reach high oil uptake capacities, optimal oil absorption selectivity and promising reusability. Such oil sorbents should present high oil uptake capacity, oil retention time and appropriate wetting behaviour. These polymeric foams are functionalized in order to increase the number of pores and sorbent capacity. Many nanocomposite polymers are used for functionalization. The functionalization is achieved by various types of coating, such as dip coating or spray coating, depending on the solubility and transparency of the materials [16]. However, the suitability of their actual use will depend on the costs associated with the production of the materials and their use and oil absorption performance. These polymeric foams can be reused more than 10 to 100 times when compared to other materials [17]. Fig. 4.3 depicts some common properties of polymeric foams.

The main property that defines the oil absorption performance is porous morphology, that is, pore size, porosity and pore connectivity.

Open porous structures, high connectivity and pore sizes about or below 500 μm can absorb oil of 40 times the polyurethane volume and can separate water from oil in a few seconds. Hence surface modification has to be carried out by defining the pore structure in order to improve efficiency [18].

4.8 Oil removal from marine environment using polymeric nanofibres

Polymeric nanofibres are sorbents that are in the form of fibres fabricated from polymer using various methods, such as electrospinning, that can immobilize oil by creating netting, thereby preventing the redispersion of oil in the environment. Nanofibres are capable of absorbing oil and expelling water due to their hydrophobic and oleophilic properties. These fibres are better fabricated using electrospinning methods because they can produce fibres of the desired size and shape with good flexibility for absorbing oil in water [19].

4.8.1 Properties of polymeric nanofibres

There are a few properties that discuss the efficiency of oil–water separation in a provided methodology. They are surface morphology, voids,

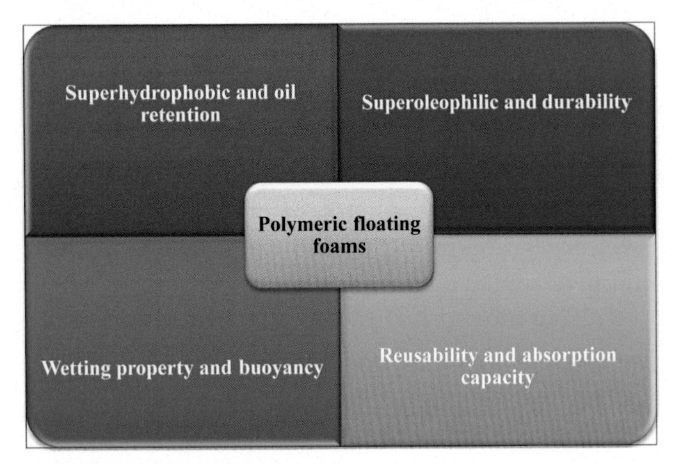

Figure 4.3 Properties of polymeric foams.

porosity, diameter of fibres and surface tension [19] . Fig. 4.4 shows most needed properties for polymeric nanofibers.

4.8.2 Mechanism of oil removal

Oil sorption in polymeric fibres is performed using three mechanisms. They are [19]:

- Adsorption
- Absorption
- Capillary action

There are certain factors that affect the efficiency of oil removal by nanofibres. Some characteristics of nanofibres are the selectivity between oil and water, specific surface area, surface roughness, buoyancy, sorbent contact angle, surface tension and spinning solution viscosity [19]. Important characteristics of oil found in water that is to be removed are viscosity, pH value, oil retention time and oil sorption kinetics. These sorbents can be reused after oil extraction using certain compression techniques like centrifugation and solvent extraction [19]. Polymeric nanofibres are an alternative choice for cleaning up oil spilled in marine water due to its versatile properties. Upgrading its production methods and surface morphology may increase its demand in the market.

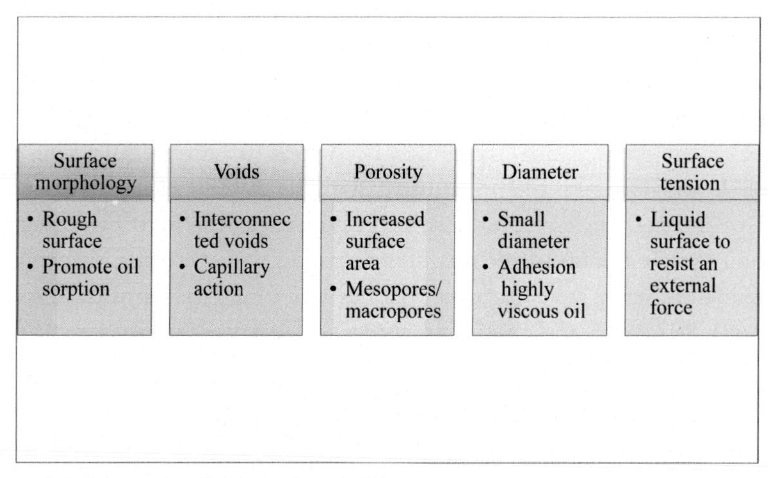

Figure 4.4 Properties of polymeric nanofibers.

4.9 Oil removal using particulate interactions

Dissolution and dispersion of oil spills in the ocean are the main reason for weathering, emulsification and mousse formation. These actions are brought about by the main factors of interactions of suspended particle and surface oil in marine water. The oil and suspended particle play a crucial role in removing oil from water. The mechanism is brought about by flocculation [20]. Turbulence due to wave action may entrain surface oil in the water column and form it into small droplets which may remain submerged and transported near to the surface by potentially adhering to oil in the surface through surface particle matter [20].

4.9.1 Role of flocculation—interaction with inorganic matters

Flocculation of fine particulate matter into larger aggregates increases the settling rate of the fines which is applied to pollutant particulates as to natural sediments. Surface particulate matter has the potential to interact with oil and cause it to sink to the seabeds. Experiments were performed with different sands, such as clay and fresh sediments, and it was found that the presence of salinity interrupts flocculation. The role of salinity also has been studied in detail, showing that the lowest rates of clay–oil flocculation are found in freshwater, higher rates in marine environments and the highest rates at lower to intermediate salinity ranges. There are various mechanisms for flocculation in inorganic matter with oil. They are adsorption of oil onto suspended particle matter, oil attaches to particles as globules and adherence of particles to oil droplets, preventing further coalescence of the oil, and thus stabilizing the suspension at sediment concentrations up to 100 mg/L. Above this the suspension is destabilized and settles down. This process – "armouring" of individual oil droplets by fine clays – has been used to prevent spilled oil from adhering to shorelines and to enhance bioremediation. When sedimented clay–oil flocs reach the depositional area further degradation of oil takes place but at a slower rate. Also there is finding stating that oil–particle aggregates transported offshore have minimum toxicity due to the extent of their dispersion [20].

4.9.2 Biological flocculation in clearing oil spills

The flocculation process involves interaction between surface oil and inorganic sedimentary materials. There are chances for oil to interact with organic materials of biological origin. Humic and fulvic acids, mucopolysaccharides and proteins are hydrophobic substances with adsorptive capacities for hydrocarbons. There are various statements suggesting that the formation of flocculation with inorganic material and oil involves incorporation of amorphous organic matter into oil slicks. Hence the formation of flocculation may be the incorporation of biogenic compounds that are found in the marine environment [20].

4.9.3 Potential of suspended particle matter to increase settling rate of surface oil

There are numerous variables involved in which oil slicks are transported to the sediments. The degree of these variables and their effects are difficult to predict. Some of the findings state that there is an interaction of oil with phytoplankton and zooplankton to remove floating oil. Sedimenting flocs of senescent cells of phytoplankton may be responsible for removing particulate oil from surface water. Ingestion of oil slicks by zooplankton and packaging in faecal pellets plays a role in removing surface oil. Physical variables, such as salinity, temperature and turbulence, are especially important in determining the degree to which surface oil and suspended particulate matter (SPM) will interact [20].

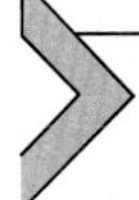

4.10 Chemical treatment using dispersants and emulsion breakers

Chemical dispersants are used under adverse weather conditions or deep water. Dispersants were used on the Deepwater Horizon oil spill in unprecedented amounts. Dispersants have two components: surfactants and solvent. When a dispersant is sprayed onto an oil slick, the interfacial tension between the oil and water is reduced, promoting the formation of finely dispersed oil droplets. There is evidence that a combination of emulsified oil and dispersant could be more toxic than oil. Hence efforts are made with dispersant formulation to make them less toxic and more biodegradable. Dispersants can be effective with viscous oils on shorelines because the contact time is prolonged, allowing better penetration of the dispersant into the oil [4].

Dispersants are specifically used in places like weathering, rough sea were skimmers cannot be applied. Its application depends on composition of oil that contaminated sea and availability. Deployment of chemical dispersants in the Arctic depends on the results of toxicity tests of chemically dispersed oil at realistic concentrations and exposures using representative Arctic species. Generally it is considered that emulsion breakers are used to recover released oil from the marine environment. The effective use of emulsion breakers depends greatly on oil properties, environmental conditions, application methods and time after a spill. Application of emulsion breakers to oil–water separators reduces the quantity of water collected, thereby improving oil collection efficiency [4].

4.11 Thermal treatment

This is the process of heating the substance at a high temperature. This is one of the advanced treatments used in oil spill removal. It involves some of the methods like incinerators and thermal desorption [4].

4.11.1 Incineration

Incineration is the high-temperature thermal oxidation of contaminants to destruction. Incinerators are found to use a variety of technologies—rotary kiln, fluidized bed and infra-red. Incinerators consist of waste storage, preparation and feeding; combustion chamber(s); air pollution control; residue and ash handling; and process monitoring. They are generally employed for soils that are contaminated with oil. They show removal over 99%. The most commonly used incinerators are rotary kiln and fixed hearth and fluidized bed. Rotary kiln and fixed hearth are twin chamber processes. The primary chamber volatilizes the organic components of the soil, and some of them oxidize to form carbon dioxide and water vapour at 650°C–1250°C. In the second chamber, high temperature oxidation (about 1100°C–1400°C) is used to completely convert the organics to carbon dioxide and water [4].

Fluidized bed incinerators are single-chamber systems containing fluidizing sand and a headspace above the bed. Fluidization with pressurized air creates high turbulence and enhances volatilization and combustion of the organics in contaminated soil [4]. Though it shows good efficiency there are some drawbacks, for example, it is a costly, high energy operation, does not fall under sustainable technology, there is the possibility of volatile metals emissions in the flue gas, and there is poor public perception due to de novo synthesis of dioxins and furans [4].

4.11.2 Low-temperature thermal desorption

This involves two process: transfer of contaminants from the soil into the vapour phase (volatilization) (about 120–600 °C); and higher temperature off-gas treatment (up to 1400 °C). It can be used for small-scale projects, as it is very flexible in operation. Unlike incineration it does not destroythe soil. It can removal petroleum hydrocarbon from all types of soil. The major drawback is formation of particulates. All low-temperature thermal desorption systems require treatment of the offgas to remove particulates and organic contaminants. Dust and organic matter can affect the efficacy of capture and treatment. The energy efficiency and economic performance of thermal desorption capacity in wet soil can be improved by pretreatment using microwave heating to remove moisture and a proportion of petroleum contamination. Microwave energy can be effectively used for the recovery of crude oil from soil. It is reported that it can recover nearly 94% of crude oil contaminants [4].

4.12 Stabilization/solidification

This is the process of immobilization of contaminants by forming solid mass, called solidification, and to prevent leaching of contaminates by stabilizing them using binders or treatment agents. Binders like lime, cement and fly ash have been used. They can be used by injecting binders into a contaminated zone. Significant reductions in total concentrations and leaching of petroleum hydrocarbons have been reported with the simultaneous improvement in soil strength due to binder addition. This is due to the combined effect of volatilization and encapsulation within the treated matrix that reduces the extractability of petroleum hydrocarbons [4].

4.13 Soil vapour extraction

It is more effective for lighter oil fraction particularly in warmer climate. It is applied to volatile compounds with a Henry's law constant greater than 0.01 or a vapour pressure greater than 0.5 mm. Crude oil has a low rate of evaporation, hence it results in low recoveries. Soil vapour extraction (SVE) removes volatile and semivolatile contaminants from the unsaturated zone by applying a vacuum connected to a series of wells. The use of a vacuum pump is to induce pressure gradient. These systems can be combined with groundwater pumping wells to remediate soil previously beneath the water table. One of the drawbacks is that it does not remove heavy oil fractions from soil, instead it encourages aerobic biodegradation [4].

4.14 Miscellaneous technologies in removing oil spills from water

Recently oil spills in marine water have been removed using wood pulp balls. Cellulose has been fabricated into balls and tested for removing oil spills from the environment. It showed good efficiency. Cellulose balls are naturally available and are biodegradable. They remove water by adsorbing oil. These balls can be effectively used up to 16 times.

4.15 Conclusion

This chapter covers lots of methods that are available. In the oil spill clean-up and remediation processes, the choice of the best technology is based on time-effectiveness, cost-effectiveness, low energy for oil recovery, good mobility, capacity to remove oil, recoverability, reusability, scaled-up production and environment-friendly characteristics. The use of material to address this environmental issue is still at the lab scale. It has to be tested at a large scale for recovering marine water from these harmful pollutants.

References

[1] Onwurah INE, Ogugua VN, Onyike NB, Ochonogor AE, Otitoju OF. Crude oil spills in the environment, effects and some innovative clean-up biotechnologies. Int J Env Res 2007;1(4):307–20.

[2] Stankovich K, Simeonova A. Techniques of cleaning up oil spills from contaminated beaches. J Sustain Dev 2018;2:29–36.

[3] Jammal NA, Juzsakova T. Review on the effectiveness of adsorbent materials in oil spills clean up. In: 7th International Conference of ICEEE, 17–19 of November 2016; Budapest, Hungary.
[4] Ivshina IB, Kuyukina MS, Krivoruchko AV, Elkin AA, Makarov SO, Cunningham CJ, et al. Oil spill problems and sustainable response strategies through new technologies. Environ Sci Process Impacts 2015;17(7):1201–19. Available from: https://doi.org/10.1039/c5em00070j.
[5] Vocciante M, Finocchi A, D'Auris ADF, Conte A, Tonziello J, Pola A, et al. Enhanced oil spill remediation by adsorption with interlinked multilayered graphene. Materials 2019;12(14):2231. Available from: https://doi.org/10.3390/ma12142231.
[6] Escudero-Oñate C, Martínez-Francés E. A review of chitosan-based materials for the removal of organic pollution from water and bioaugmentation. Chitin-Chitosan: Myriad Functionalities in Science and Technology; 2018. doi:10.5772/intechopen.76540.
[7] Lutfee T, Al-Najar JA, Abdulla FM. Removal of oil from produced water using biosorbent. In: IOP Conference Series: Materials Science and Engineering; 2020; 737: 012198. doi:10.1088/1757-899X/737/1/012198.
[8] Ghosal D, Ghosh S, Dutta TK, Ahn Y. Current state of knowledge in microbial degradation of polycyclic aromatic hydrocarbons (PAHs): a review. Front Microbiol 2016;7. Available from: https://doi.org/10.3389/fmicb.2016.01369.
[9] Hassanshahian M, Abarian M, Cappello S. Biodegradation of aromatic compounds. Biodegradation and Bioremediation of Polluted Systems - New Advances and Technologies; 2015. doi:10.5772/60894.
[10] Peng R-H, Xiong A-S, Xue Y, Fu X-Y, Gao F, Zhao W, et al. Microbial biodegradation of polyaromatic hydrocarbons. FEMS Microbiol Rev 2008;32(6):927–55. Available from: https://doi.org/10.1111/j.1574-6976.2008.00127.x.
[11] Karlapudi AP, Venkateswarulu TC, Tammineedi J, Kanumuri L, Ravuru BK, Dirisala VR, et al. Role of biosurfactants in bioremediation of oil pollution-a review. Petroleum 2018;4(3):241–9. Available from: https://doi.org/10.1016/j.petlm.2018.03.007.
[12] Silva R, Almeida D, Rufino R, Luna J, Santos V, Sarubbo L. Applications of biosurfactants in the petroleum industry and the remediation of oil spills. Int J Mol Sci 2014;15(7):12523–42. Available from: https://doi.org/10.3390/ijms150712523.
[13] Calcagnile P, Fragouli D, Bayer IS, Anyfantis GC, Martiradonna L, Cozzoli PD, et al. Magnetically driven floating foams for the removal of oil contaminants from water. ACS Nano 2012;6(6):5413–19. Available from: https://doi.org/10.1021/nn3012948.
[14] Peng L, Yuan S, Yan G, Yu P, Luo Y. Hydrophobic sponge for spilled oil absorption. J Appl Polym Sci 2014;131(20). Available from: https://doi.org/10.1002/app.40886.
[15] Zhu K, Shang Y-Y, Sun P-Z, Li Z, Li X-M, Wei JQ, et al. Oil spill cleanup from sea water by carbon nanotube sponges. Front Mater Sci 2013;7(2):170–6. Available from: https://doi.org/10.1007/s11706-013-0200-1.
[16] Barroso-Solares S, Pinto J, Fragouli D, Athanassiou A. Facile oil removal from water-in-oil stable emulsions using PU foams. Materials 2018;11(12):2382. Available from: https://doi.org/10.3390/ma11122382.
[17] Pinto J, Athanassiou A, Fragouli D. Surface modification of polymeric foams for oil spills remediation. J Environ Manag 2018;206:872–89. Available from: https://doi.org/10.1016/j.jenvman.2017.11.060.
[18] Pinto J, Athanassiou A, Fragouli D. Effect of the porous structure of polymer foams on the remediation of oil spills. J Phys D: Appl Phys 2016;49(14):145601. Available from: https://doi.org/10.1088/0022-3727/49/14/145601.

[19] Sarbatly R, Krishnaiah D, Kamin Z. A review of polymer nanofibres by electrospinning and their application in oil–water separation for cleaning up marine oil spills. Mar Pollut Bull 2016;106(1-2):8–16. Available from: https://doi.org/10.1016/j.marpolbul.2016.03.037.
[20] Muschenheim D, Lee K. Removal of oil from the sea surface through particulate interactions: review and prospectus. Spill Sci Technol Bull 2002;8(1):9–18. Available from: https://doi.org/10.1016/s1353-2561(02)00129-9.

CHAPTER FIVE

Ballast water management

5.1 Introduction to ballast water

Ballast water is water providing stability and trims to ships that are sailing in marine water. This water is in tanks provided in ships known as ballast tanks. When the ship is loaded with cargo it has weight, hence ballast water is not important, whereas when a ship discharges cargo it is not carrying weight and hence ballast water is needed. It is an essential component to ensure a vessel's stability, structural integrity and safe navigation. It allows a vessel to adjust for the changes that occur in port. Ballast water contains many marine microorganisms, phytoplankton, zooplanktons, which when released to the marine environment may possess serious threats to local marine life/biota [1]. Such invasive aquatic species are serious threats to the marine ecosystem. Some of the impacts are categorized as affecting ecological, economical and/or human health. The division in impacts is provided by the International Maritime Organisation (IMO). Identification of these problems by governmental and nongovernmental organization, industries and the global scientific bodies promoted changes in regulating ballast water. IMO adopted the international convention for the control and management of ships' ballast water and sediment which came into force in 2017. The main agenda of this convention is to prevent, minimize and eradicate the transfer of harmful aquatic organisms and pathogens by controlling and managing ballast water from ships [1,2]. There are many options used in preventing aquatic species entering ballast water using either chemical or physical treatments.

5.2 Ballast water quality by international maritime organisation

Discharge of ballast water has to follow the regulation imposed by IMO, which is called regulation D-2 ballast water performance standard.

Modern Treatment Strategies for Marine Pollution.
DOI: https://doi.org/10.1016/B978-0-12-822279-9.00011-7

Table 5.1 Regulation of discharged ballast water.

Organism	Regulation of discharged ballast water	Reference
Phytoplankton/zooplankton ≥ 50 μm	<10 viable organisms per m^3	[1]
Phytoplankton/zooplankton 10–50 μm	<10 viable organisms per mL	
Toxicogenic *Vibrio cholera* (O1 and O139)	<1 colony forming unit per 100 mL	
Escherichia coli	<250 colony forming unit per 100 mL	
Intestinal enterococci	<100 colony forming unit per 100 mL	

This regulation includes two indicators concerning the size of the plankton organisms, and three indicators concerning human health standards [1]. Table 5.1 shows the regulation of discharged ballast water.

5.3 Regulations for ballast waste management

5.3.1 International regime

At international level ballast water and its impacts are addressed by IMO and its subsidiary body Marine Environment Protection Committee (MEPC). IMO designed the Ballast water management convention, along with discharge standards which is discussed in Table 5.1. After further review and a revision process regulation D-2 was changed into D-3. This addressed issues like the use of active substances like biocidal agents, to provide environmental and safety measures, referred to as the G9 guidelines [3].

5.3.2 US federal domestic regime

US provides ballast water management by two statutes:

1. Nonindigenous Aquatic Nuisance Prevention and Control Act (NANPCA) of 1990.
2. Federal Water Pollution Control Act (commonly referred to as the Clean Water Act or CWA).

In 1990 a US Congressional Act directed the US Coast Guard (USCG) to issue regulations to limit ANS into the Great Lakes from ballast water.

USCG issues final mandatory ballast water management (BWM) regulations for the Great Lakes. USCG issued rules to establish national voluntary BWM program. In 2004 the USCG issued rules making the national program mandatory. Finally in 2012 the USCG released ballast water (BW) discharge limits and a type approval process for BW treatment equipment [3]. In 2010 the Environmental Protection Agency (EPA) commissioned scientific studies to inform BW requirements for vessel general permits. In 2011 the EPA published new draft vessel general permits with discharge limits [3].

5.4 Ballast water treatment

According to the IMO guidelines there are lots of best measures to reduce the risk of transfer of harmful aquatic organisms. There are a number of techniques that can be used to minimize the introduction of nonindigenous aquatic nuisance species from entering into ballast water. There is no single technique to treat ballast water but a combination of technology is effective in removing invasive species from water. As it is very difficult to erect ballast water technology because it is costly, unreliable and time consuming, there are some criteria available to choose from [1]. Fig. 5.1 shows different methods for BWM. They are:

- safety of the crew and passengers;
- effectiveness in removing target organisms;

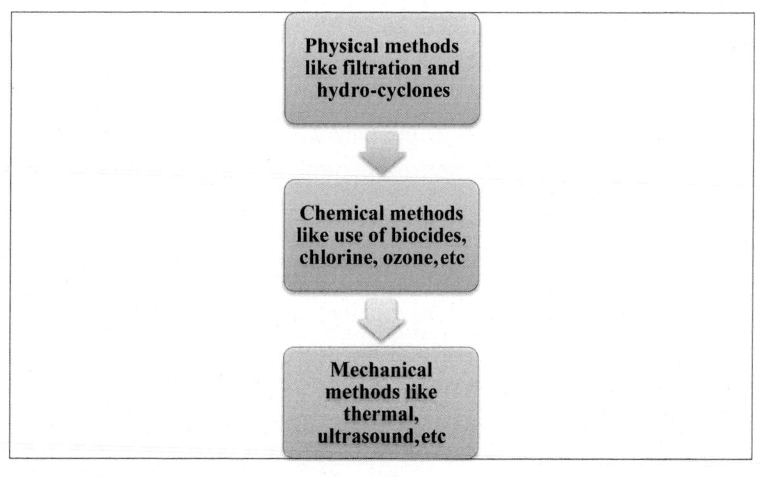

Figure 5.1 Different types of methods for BWM.

- ease of operating treatment equipment;
- amount of interference with normal ship operations and travel time;
- structural integrity of the ship;
- size and expense of treatment equipment;
- amount of potential damage of the environment; and
- ease of port authorities to monitor for compliance with regulations [1].

5.4.1 Primary treatment

Primary treatment denotes physical methods like filtration and hydrocyclone. Filtration is the most effective method for treating ballast water. It helps in screening solid particles of various sizes that can be prevented from entering into sea water. Whereas hydro cyclone is a cyclonic separation that requires a lower pressure pump and allows the separation of sediments and suspended solids of size 20 μm. A multilevel filter technique and a cyclonic separator followed by ultraviolet radiation was used for removing various organisms like zooplanktons, phytoplankton, bacteria and viruses. It reduced macrozooplankton by over 95% and microzooplankton (rotifers) by more than 90% [1].

5.4.2 Mechanical treatment

Mechanical separation includes the use of ultraviolet radiation, heat treatment and electric pulse applications in removing invasive species from ballast water. UV radiation is commonly used for the disinfection of surface and wastewater. UV proved to be the best in killing bacteria and viruses without causing any serious hazardous by-products. UV radiation operates by causing photochemical reactions with biological components such as nucleic acids (DNA and RNA) and proteins. The effectiveness of UV treatment depends on the size and morphology of the organisms. It is considered as a secondary option for ballast waste treatment. Most of the time it is combined with filtration (screening) or the use of hydrocyclones in ballast water management [1].

Heat-thermal treatment is used with the following options: (1) the use of waste heat produced by the ship's engines; and (2) the use of heat created by backup boiler systems installed aboard the vessel. The minimum temperature needed to deactivate unwanted species is over 40°C. It can be used when ballast water enters from a warmer environment. There are some drawbacks in opting for heat treatment, like the availability of waste heat, the method of exchanging this heat, the length of the voyage, the cooling effect of passages in cold weather, the effect of the heat removal on engine performance, the potential of increased corrosion and the probability of high maintenance with the heat exchangers [1].

The use of microwaves involves the application of microwaves for heating application. They have high heating rate compared to conventional heating. Researchers have developed a continuous microwave with operational parameters like rate of flow, power level and temperature. Selected microorganisms are inactivated using this technology [1].

Small-scale experiments have been performed by applying an electrical voltage in the 15–45 kV range with a pulse duration of 1 μs in operating electric pulse technology. Large energy sources would be required for systems capable of treating large volumes of ballast water [1]. Ultrasound or sonication technology is another methodology that can be used as a secondary treatment option. 1.4 kHz has been used in controlling the growth of organisms like blue green algae. Green algae removal was 40% for a residence time of 20 s [1]. Fig. 5.2 shows some of the physical and mechanical treatments in BWM.

5.4.3 Chemical treatment

There are a wide variety of chemicals used for cleaning ballast water. Some of them are discussed below.

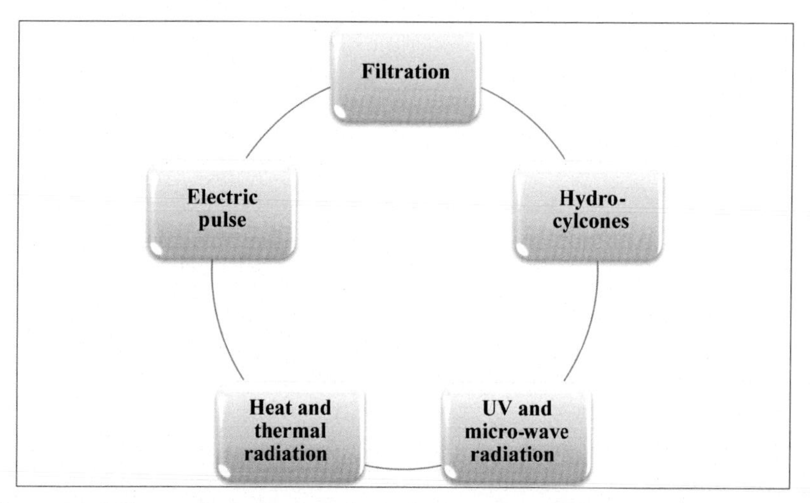

Figure 5.2 Different physical and mechanical treatments in ballast water management.

5.4.3.1 Biocides

The addition of chemicals that would kill or inactivate a variety of organisms found in ballast water is an attractive treatment technology. A biocide is added to ballast tanks and allowed to react for some time. The principal mechanisms through which organisms are killed are damage to the cell wall, alteration of cell permeability, alteration of the colloidal nature of the protoplasm, alteration of the organism DNA or RNA and inhibition of enzyme activity [1]. They are classified into two groups.

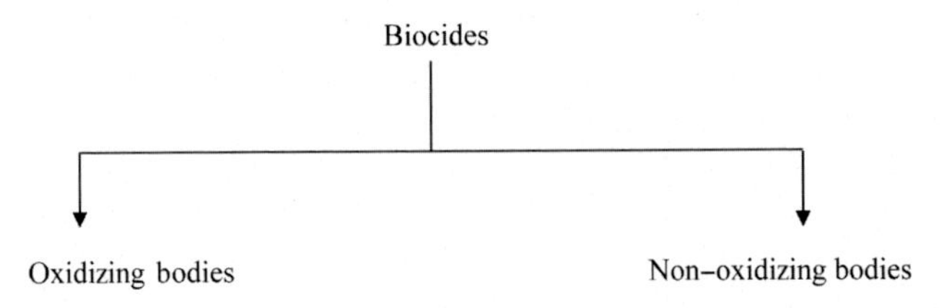

Oxidizing bodies are used in freshwater because the water is depleted by organic matter. They include chlorine, chlorine dioxide, ozone, bromine, hydrogen peroxide and peroxy acetic acid. Nonoxidizing bodies are chemical compounds that are dibromonitrilo-propionamide, formaldehyde, glutaraldehyde, quaternary ammonium salts, various organosulfur compounds, anionic and nonanionic surface-active agents [1].

5.4.3.2 Oxidizing biocides

5.4.3.2.1 Chlorine

Chlorine is a strong oxidizing agent. Its effectiveness depends on temperature, residual chlorine and reaction time. Before discharging ballast water it should be dechlorinated by using sulfur dioxide which reacts with residual chlorine to form chlorine ions [1].

5.4.3.2.2 Chlorine dioxide

It is costly and produces by-products. It is also a strong disinfectant. Chlorine dioxide is toxic to aquatic organisms, but under normal dosing conditions it will decline to very low levels before release [1]. Chlorite and chlorate ions are two disinfection by-products formed as toxic by-products.

5.4.3.2.3 Ozone

Ozone is a very powerful but unstable, oxidizing agent which destroys viruses and bacteria, including spores when used as a disinfectant in water treatment plants. There is potential for ozone to remove the marine dinoflagellate algae, *Amphidinium* sp. but the removal of this species from ballast water requires high dosages of ozone [1]. Disinfection by-products are formed when oxidizing disinfectants with organic and inorganic matter present in water that pose serious environmental or health hazards. One of the products is trihalomethanes as a result of the reaction between bromine and natural organic matter. There are two categories of ozone disinfection by-products (DBPs): nonbrominated (aldehydes, acids, aldo- and ketoacids) and brominated compounds. The presence of bromide ion in seawater enhances the production of bromate ion and bromoform [1].

5.4.4 Electroionization magnetic separation

This involves several sequential processes as follows [1]:

- continuous flow of ionized gas (containing oxygen and nitrogen) into the water;
- air is passed through strong ultraviolet and magnetic fields thereby creating oxygen and nitrogen ions;
- injection of ions in to water;
- coagulation of contaminates in water;
- flocs are formed by flocculation and separated by magnetic separation filtration; and
- they remove phytoplankton with 92% efficiency [1].

5.4.5 Deoxygenation

This involves bubbling nitrogen or other inert gas into ballast water to reduce its oxygen content. Deoxygenation treatment technologies involve the use of glucose, carbon monoxide, or bioreactors containing fixed beds of oxygen-removing bacteria that are all designed to deplete oxygen levels from seawater. They have a protective effect for pipework but requires 1–4 days to gain optimum effectiveness [1].

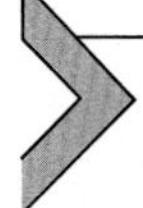

5.5 Ballast water management system using active substances

A variety of ballast water management system (BWMS) using physical and chemical technologies have been developed over the years. The most common technique is in situ electrolysis (electrochlorination) where seawater is allowed to produce its own oxidizing derivatives, termed as total residual oxidants (TRO). This process is combined with filtration or other separation steps followed by neutralization step before discharge [4].

5.5.1 Electrochlorination

In this process seawater is allowed to produce the active substance TRO. It is commonly found in ships for treating ballast water contaminants. In BWMS, the electrolysis unit generating the active substance is either mounted directly in the main ballast water pipeline to produce concentrated active substance or injected into the ballast water pipelines. If ships have to operate in freshwater and have an in situ electrolysis system on board, then these ships have to use a separate brine tank as source water for BWMS [5].

5.5.1.1 Reaction mechanism

Oxidants used for generating active substances in seawater are hypobromous acid and hypobromite ($HOBr/BrO^-$), and hypochlorous acid and hypochlorite ($HOCl/ClO^-$), which are strong oxidants of organic matter including aquatic organisms. When seawater, which act as the electrolyte, has a direct current passed across two electrodes, a chemical reaction takes place where hydrogen and chlorine gas are formed; much of the chlorine gas dissolves in the water [5].

At anode [5]

$$2Cl^- \rightarrow Cl_2 + 2e^-$$

At cathode [5]

$$2Na^+ + 2e^- + 2H_2O \rightarrow 2NaOH + H_2$$

In electrolyte solution [5]

$$Cl_2 + 2NaOH \rightarrow NaClO + NaCl + H_2O$$

$$NaClO + H_2O \rightarrow HClO + Na^+ + OH^-$$

$$HClO \rightarrow H^+ + ClO^-$$

5.5.1.2 Chlorine gas

Chlorine gas and water reacts to form hypochlorous acid (HOCl) and hydrochlorite ions (ClO^-). Hypochlorous acid is a weak acid (pKa of about 7.5), and dissociates into hydrogen and hypochlorite ions [5].

$$Cl_2 + H_2O \rightarrow HOCl + HCl$$
$$HOCl \rightarrow H^+ + ClO^-$$

The ClO^- and HOCl species are commonly referred to as free active chlorine (FAC), which is extremely reactive with the cell components of numerous microorganisms [5]. There are other TRO disinfectants used for cleaning microbes. They are peracetic acid, hydrogen peroxide (H_2O_2), ozone (O_3), hydroxyl radicals (•OH), chlorine dioxide (ClO_2) and sodium dichloroisocyanurate (NaDCC) [5].

5.5.2 Ozonation

Ozone is an unstable gas, consisting of three oxygen atoms which readily degrade back to oxygen. During this transition a free oxygen atom, or free radical, is formed which is highly reactive and short-lived, under normal conditions. Due to this high oxidation potential ozone oxidizes cell components of the cell wall like RNA, DNA, enzymes and proteins. As they are strong oxidants they are used for oxidizing bromide in seawater to bromate [5].

$$O_3 + Br^- \rightarrow O_2 + BrO^-$$
$$2O_3 + BrO^- \rightarrow 2O_2 + BrO_3^-$$

5.5.3 UV light

The main mode of action of UV light is to damage the genetic material of the organisms, thus making them unable to reproduce. UV affects all organisms but has the highest impact on organisms in the 0–50 μm range, mainly algae, due to the size-related penetration depth [5].

5.5.4 Neutralization of active substance

BWMS is equipped with a neutralization step to ensure compliance with the maximum allowable discharge concentration (MADC) of active substance. Treatment with reducing agents like sodium thiosulfate or sodium bisulfite is applied. In this way excess oxidants are not released into the water [5].

$$4NaClO + Na_2S_2O_3 + 2NaOH \rightarrow 4NaCl + 2Na_2SO_4 + H_2O$$

5.6 Ballast water exchange

This method has been opted for by various countries. There are various types, which are sequential or reballasting and flow through or continuous flushing. Sequential ballast water exchange involves completely emptying ballast tanks and refilling them with ocean water. Continuous ballast water exchange involves partially emptying and refilling the tanks. The effectiveness of these methods depends on the piping system fitted in ships. Additionally the structure of the ship plays a crucial role [1].

5.6.1 Sequential ballast water exchange

This method involves emptying and refilling ballast tanks. The emptying and refilling procedure is commonly accomplished by using the existing water intake/suction piping system and ballasting pumps which are used to empty and refill the tanks. According to the IMO guidelines, ballast water is to be discharged until suction is lost in the ballast tanks. It also suggested that during full tank ballasting procedures, large changes in loading conditions could affect stability, strength, draft and trim of the ship. Emptying and refilling tanks will not remove all the sediments and organisms at the bottom of the ballast tanks. Continuous empty and refilling is suggested for releasing all particles from tanks [1].

5.6.2 Flow through ballast water exchange

This method involves ballast tanks are flushed out by pumping mid-ocean water into the ballast tanks during a voyage allowing port water to overflow out. It requires a separate uptake and outflow system. Water is poured into tanks through common water intakes originating from the sea chest. According to IMO guidelines tanks should be pumped out three times [1].

5.7 Problems associated with ballast water

Ballast water is an environment with several habitats. They not only carry organisms as plankton in water, but in addition the tank bottom, with accumulated sediment, and the tank walls provide space for species to settle and flourish [6].

5.7.1 Sediments in ballast tanks

Some species may either settle to the tank bottom by gravity or may complete their meroplanktonic life cycle inside the tank and, in the absence of other habitats, they colonize tank bottoms with or without sediment layers. According to IMO guidelines, sediment settling should be prevented and growth of organisms should be avoided for easy removal of settled sediments from the bottom of the tanks. Initially sediments settle to the tank bottom and begin to accumulate. The amount of accumulated sediment in a tank is influenced primarily by the volume of suspended matter in the water from which the ship draws its ballast, by the ship's ballast management practices, the type and design of ballast tanks, and the time since the tanks were cleaned since last in a dry dock [6].

5.7.2 Biofouling in the ballast tanks

Some species may complete their life cycle in tanks by settling to the tank walls. This kind of biofouling in the tank may spread species when they spawn between periods of tank wall cleaning. Through this way the organisms are introduced into the new environment [6].

5.7.3 Larger organism in the tank

As the ballast water supply system has manholes there is the possibility for larger fish of 10 cm length to swim inside the tanks. Larger organisms growing inside these tanks are observed in some ballast water supplies [6].

5.7.4 Trap samples

They are used to retrieve samples without emptying the tanks. The traps are made from poly vinyl chloride (PVC) cylinders of 50 cm length with an internal diameter of 16.8 cm. Funnel-shaped steel meshes (0.8 mm mesh size) with a median opening of 5 cm cover both openings of each cylinder. Species caught in traps are juvenile fish, small shrimps and juvenile decapods, none of which were found in net samples with nets operated through the manhole of the same ballast tank. Traps with light and bait as attractant captured more specimens and taxa than the empty control traps [6]. Fig. 5.3 depicts overview on ballast water treatment protocol.

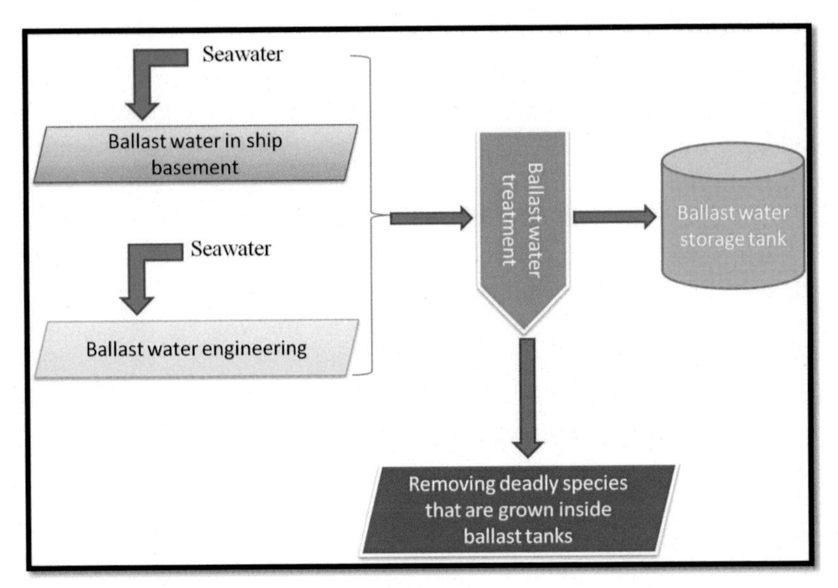

Figure 5.3 Overview of ballast water treatment.

5.8 Conclusion

This chapter precisely covers the treatment technologies followed in cleaning ballast tanks. All potential methods of ballast water treatment involve an operational cost which is of major concern for ship owners and quite often the most important parameter for choosing a ballast water treatment technology. All treatment requires proper cleaning to avoid fouling effects. The main point to be emphasized in the consideration of the application of each method is the impact that the ballast water treatment system might have on the new marine environment. Toxicity is a parameter that must be evaluated before choosing ballast technology. Added substances in the marine environment must not deliver toxicity effects over the whole marine life range. While several ballast water treatment technologies have been certified according to the IMO guidelines, further evaluation is necessary with regard to the development of new processes at lab, pilot and full scale, as well as study of the environmental implications of these technologies.

References

[1] Tsolaki E, Diamadopoulos E. Technologies for ballast water treatment: a review. J Chem Technol Biotechnol 2010;85(1):19–32. Available from: https://doi.org/10.1002/jctb.2276.

[2] Estévez-Calvar N, Gambardella C, Miraglia F, Pavanello G, Greco G, Faimali M, et al. Potential use of an ultrasound antifouling technology as a ballast water treatment system. J Sea Res 2018;133:115–23. Available from: https://doi.org/10.1016/j.seares.2017.04.007.

[3] Albert RJ, Lishman MJ, Saxena RJ. Ballast water regulations and the move toward concentration-based numeric discharge limits. Ecol Appl 2013;23(2):289–300.

[4] www.unesco.org <http://www.gesamp.org/publications>

[5] GESAMP (IMO/FAO/UNESCO-IOC/UNIDO/WMO/IAEA/UN/UN Environment/UNDP/ISA Joint Group of Experts on the Scientific Aspects of Marine Environmental Protection); 2019. Methodology for the evaluation of ballast water management systems using Active Substances. Rep. Stud. GESAMP No. 101, p. 110.

[6] Gollasch S, David M. Ballast water: problems and management. World Seas 2019;237–50. Available from: https://doi.org/10.1016/b978-0-12-805052-1.00014-0.

CHAPTER SIX

Pesticides clean-up

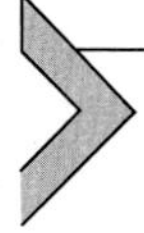

6.1 Introduction to pesticides pollution in marine environment

Pesticides are substances or mixtures of substances designed to control, kill, or regulate the growth of undesirable biological organisms. These undesirable biological organisms (pests) do not only compete with humans for food and transmit diseases but they are generally a nuisance to the environment. Pests include insects, plant pathogens, weeds, mollusks, fish, birds, mammals, nematodes and microorganisms such as bacteria and viruses. Pesticides may be classified as being biological or synthetic. Biological pesticides are derived from natural sources like plants and their extracts and microorganisms. Synthetic pesticides are chemically made by man and are driven by external factors. There is further classification of pesticides like broad spectrum and narrow spectrum pesticides. Broad spectrum pesticides are used to control a wide range of species, whereas narrow spectrum pesticides target a certain group of pests. Pesticides include insecticides (control insects), herbicides (control weeds) and fungicides (control fungi) [1].

Pesticide pollution may affect both aquatic and soil ecosystems. There are numerous factors that promote pesticide pollution. They are drainage patterns, properties of the pesticide, rainfall, microbial activity, treatment surface and rate of application. Pesticides are transported into aquatic systems either as single chemicals or complex mixtures through various processes, such as direct applications, surface runoffs, spray drifts, agricultural returns and groundwater intrusions. Toxic pesticides are responsible for the death of fish and zooplankton. There are various factors that affect the toxic nature of pesticides. They are chemical structure, water solubility and polarity, volatility and the ingredients used for the formulation [1].

There are lots of acute effects on fish, amphibia, crustacea and birds after the use of dichlorodiphenyltrichloroethane (DDT) for insect control in aquatic zones. There are a number of reports stating increased mortalities in marine fishes and biota due to the application of pesticides [2]. There are

Modern Treatment Strategies for Marine Pollution.
DOI: https://doi.org/10.1016/B978-0-12-822279-9.00010-5

various metabolisms that favour the change in living organisms due to the chemical structure of pesticides. It was suggested that fish are lacking in the ability to metabolize foreign compounds, though they have an oxidase enzyme which is responsible for the oxidation of organic compounds. The occurrence of organic chemicals in the marine environment requires an evaluation of their effects, whether directly on a species or indirectly on the marine ecosystem. In aquatic toxicology, the acute toxicity test for invertebrates and fish enables an estimation of the exposure concentration resulting in 50% mortality of the test animals within 48 or 96 h, which is called LC_{50}. There are many factors that affect the LC_{50} values of organic chemicals to fishes. They are species relationships, temperature, salinity, oxygen concentration, developmental stage, physiological condition, time of exposure and experimental conditions such as continuous flow or static exposure. All these effects are possible for the release of pollutants into the sea and depend on depth and location aspects [3].

These pesticides affect organisms living in the natural environment, thereby changing the quality and quantity of the population. The marine or aquatic organisms can make use of substances entering the environment as feeding substrate for a source of energy or a building material, thus affecting the balance in the ecosystem. There are numerous studies stating that effects are witnessed in the littoral zone between the terrestrial and aquatic environment. It is a hotspot for many aquatic and terrestrial bacteria and fungi and the place where energy transformation takes place. This zone act as a filter, thereby protecting the marine ecosystem from being affected by harmful substances. A high concentration of organic substances and good thermal–aerobic conditions make this zone a place where autotrophic and heterotrophic bacteria and fungi can flourish, hence there are chances for microorganisms to be affected by pesticides entering [4]. There are lot of pesticides that are reported in marine biota as residues. They are DDT, chlordane, dieldrin, endrin and many more [5]. This chapter covers some of the treatment methodologies that are used to remove pesticides from the aquatic environment.

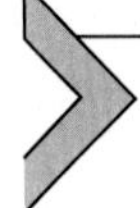

6.2 Removal of pesticides from marine water using different methodologies

Due to the growth of pesticide pollution in marine water, effective cleanup is needed for any particular polluted site in order to prevent harm

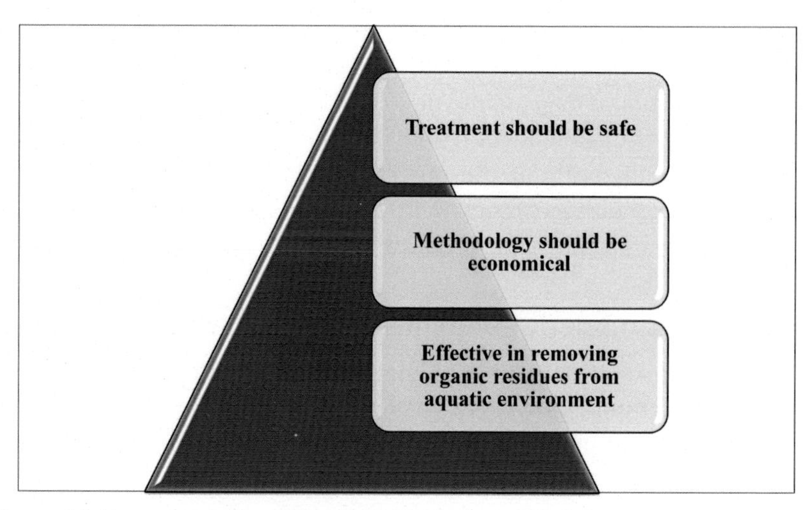

Figure 6.1 Strategies to be adopted satisfied before choosing treatment.

to aquatic microorganisms. The technology adopted for cleaning of environment should be safe, economical and efficient. It should be able to be used in all possible conditions. Some technologies include phytoremediation, microbial degradation, photodegradation, membrane removal and thermal desorption. The most commonly employed method is microbial and photodegradation. This chapter covers in detail the photo- and microbial degradation of pesticides and their residues that are present in the aquatic environment. Fig. 6.1 gives strategies to be adopted while choosing treatment methodology for pesticide cleaning.

6.3 Microbial degradation of pesticides in aquatic environment

There are certain physical treatments, such as adsorption and percolator filters, and chemical treatments, such as the advanced oxidation, which involve the use of powerful transient species, mainly the hydroxyl radical which involves the formation of toxic by-products. Also it is not cost-effective. Hence the development of microbial degradation took place. Bioremediation is an innovative technology that is being used for the clean-up of polluted sites. This technology is cost effective and becoming an increasingly attractive clean-up technology. The process can be carried out by using indigenous microorganisms or by adding an enriched culture of microorganisms. Microbes utilize their inherent specific characteristics to degrade the desired contaminant at a faster rate. The result of bioremediation is the complete mineralization of contaminants to H_2O and CO_2 without the build-up of intermediates. For effective bioremediation, microorganisms must enzymatically attack the pollutants and convert them to less toxic products [6].

6.3.1 Various modes of bioremediation with microorganisms

Basically there are three types of bioremediation with microorganisms: remediation through improved natural attenuation, bioaugmentation (introduction of nonnative or genetically modified organisms) and biostimulation (addition of an electron acceptor or nutrients) [7].

6.3.1.1 Strategies in accessing bioaugmentation

There are some key factors in accessing the success of bioaugmentation. They are [7]:

- Ecological basis, in terms of acquiring basic knowledge about the microbial metabolic processes.
- Monitoring techniques: the main techniques in this field are based on polymerase chain reaction (PCR) to determine the activity and/or survival of added microorganisms.
- Plant management, in terms of anticipating the seasonal effects on microbial and/or consortium communities used in bioaugmentation.
- Selection criteria.
- Single inoculation versus continuous application in the environment.
- Immobilization techniques that are used to increase the metabolic activity of microorganisms.
- Gene transfer.
- Membrane reactors with the use of start-up degraders to accelerate the reaction.

6.3.1.2 Comparison of efficiencies of bioremediation in all three types

The efficiency of bioremediation by natural attenuation, bioaugmentation and biostimulation was compared using carbofuran. In natural attenuation they found that the compound was degraded with $t_{1/2}$ of 16.63 days. Use of bioaugmentation yielded degradation with $t_{1/2}$ of 1.60 days. With the help of sludge from ethanol production and methane production, hydrogen fermentation yielded better results as a biostimulating agent with $t_{1/2}$ 9.53 days [7].

6.3.2 Bacterial degradation of pesticides

The isolation and growth of microbes for degrading the pollutants is the new tool to restore the pollutant environment. Upon complete biodegradation of the pesticide, carbon dioxide and water are formed by the oxidation of the parent compound and this process provides the energy to the microbes for their metabolism. The intracellular or extracellular enzymes of the microbes play a major role in the degradation of chemical compounds. Microbes have degraded pesticides like organochlorine compounds, organophosphorous, neonicotenoids and carbamate. Some of the organochlorine pesticides include dieldrin, aldrin, alpha-BHC, beta-BHC, delta-BHC, gamma-BHC (Lindane), heptachlor, endosulfan, methoxychlor, aroclor and DDT. Table 6.1 shows some common microbes used for degradation of pesticides.

Studies showed that a wide variety of bacteria could degrade carbamates using carbofuran hydrolase. Among other genera *Pseudomonas*, *Mesorhizobium*, *Ralstonia*, *Rhodococcus*, *Ochrobactrum* and *Bacillus* are the most notorious.

Table 6.1 Microorganisms used in pesticides degradation.

Pesticides	Microorganism	References
Dichlorodiphenyltrichloroethane (DDT)	*Alcaligenes eutrophus*, *Aerobacter aerogenes*, *Sphingobacterium* sp., *Penicillium miczynskii*, *Aspergillus sydowii*, *Trichoderma* sp., *Penicillium raistrickii*, *A. sydowii*, *Bionectria* sp., *A. aerogenes*, *Trichoderma viridae*, *Pseudomonas* sp., *Micrococcus* sp., *Arthrobacter* sp., *Bacillus* sp., *Pseudomonas* sp.	[6]
Dichlorodiphenyldichloroethylene (DDE)	*Phanerochaete chrysosporium*	[6]
Dichlorodiphenyldichloroethane (DDD)	*Trichoderma* sp.	[6]
Dieldrin	*Pseudomonas* sp.	[6]
Endosulfan	*Pseudomonas aeruginosa*, *Burkholderia cepaeia*, *Arthrobacter* sp. *KW*, *Aspergillus niger*	[6]
Pentachlorophenol (PCP)	*Arthrobacter* sp., *Flavobacterium* sp.	[6]

Generally carbamates are transformed into various products through several processes like hydrolysis, biodegradation, oxidation, photolysis, biotransformation and metabolic reactions in living organisms [6]. The microbes undergo metabolic reactions based on aerobic and anaerobic conditions. Degradation of toxic compounds are favourable in anaerobic conditions due to reasons like the easy pathway and capacity of microbes to tackle the environment.

6.3.3 Fungal degradation of pesticides

When the bacterial species fails to degrade pesticides then fungus can be used. The wood attacking fungi, such as *Phanerochaete* and other related fungi, have a powerful extracellular enzyme called peroxidase that acts on a broad range of chemical compounds. A variety of chemicals including pesticides, polychlorinated biphenyl (PCB) and polycyclic aromatic hydrocarbons (PAH) are degraded by *Phanerochaete chrysosporium*. It was reported that successful degradation of endosulfan and endosulfan sulfate was achieved by a white-rot fungus *Trametes hirsuta*. The fungus showed degradation of endosulfan and endosulfan sulfate using multiple pathways. The fungal strains *Fusarium oxysporum*, *Lentinulaedodes*, *Penicillium brevicompactum* and *Lecanicillium saksenae* have great potential for the biodegradation of the pesticides including difenoconazole, terbuthylazine and pendimethalin in batch liquid cultures and are investigated to be valuable as active microorganisms for pesticides degradation [6].

6.3.4 Enzymatic degradation of pesticides using microbes

A biochemical reaction for the degradation of pesticides is achieved using different enzymes like dehydrogenases, cytochrome P450, dioxigenases and ligninases. Bacteria and fungi produces enzymes like hydrolytic enzymes, oxygenases and peroxidases that are responsible for degradation. It is found that hydrolysis is the major reaction that takes place to cleave the toxic compounds into multiple products. Other reactions like oxidation and reduction take place for the complete mineralization of these compounds [6]. The accumulation of residues in aquatic organisms is a function of the pesticide concentration to which they are exposed and the time at which they are exposed. It is therefore important to know the persistence of these compounds in the estuarine environment. Some experiments were carried out to determine the degradation and accumulation of malathion, endosulfan and fenvalerate in sea fishes. Half-lives of these compounds at varying pH vary with respect to the water condition.

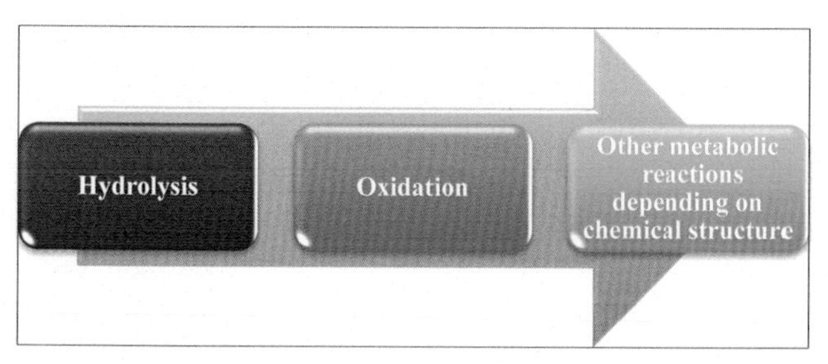

Figure 6.2 Metabolic reactions.

Half-lives vary from 22 to 12 days for each compound, inducing either accumulation or degradation. The half-lives of certain compounds, like malathion, decrease, which shows that hydroxide-catalyzed hydrolysis is a major pathway for their degradation in marine systems. Microbial action may also play a role in the degradation of malathion and fenvalerate [8]. Although enzymatic degradation achieves better results than other bioremediation techniques, it is sensitive to pH, organic solvents, different temperatures and certain environmental conditions through which the ability to degrade specific substrates may be lost [7]. Fig. 6.2 depicts some common metabolic reactions during degradation process.

6.4 Photodegradation of pesticides

Photodegradation is a kind of oxidation process using the external energy of sunlight and a catalyst like TiO_2 for the production of hydroxyl radicals to degrade organic chemical compounds. It is considered to be an advanced oxidation process depending on the applicability and use of different compounds to form free radicals. It is commonly called a near-ambient temperature treatment process based on highly reactive radicals, especially the hydroxyl radical (•OH). Other radicals and active oxygen species involved are superoxide radical anions ($O_2^{\bullet -}$), hydroperoxyl radicals ($HO_2^{\bullet -}$), triplet oxygen ($3O_2$) and organic peroxyl radicals (ROO^-) [9]. It is reported by Pandit et al. [10] that organochlorine pesticides are present in the Indian marine environment in sediment and marine fishes. These compounds accumulate in the lipid components of biological species and resist degradation. In India DDT and hexachlorocyclohexane (HCH) are used for agricultural and sanitary purposes. About 25,000 Mt of chlorinated pesticides were used in India, and DDT accounted for about 40%. Identified samples showed the extent of the contamination of marine organisms by these pesticides and the nature of accumulation that was characteristic in marine life. There are different kinds of photodegradation, such as direct, photo-sensitized and photocatalytic degradation. Direct photodegradation is the process by which sunlight directly enters the surface of water and degrades the organic compounds that are present. Direct irradiation will lead the pesticides to their excited singlet states. This excited state may undergo homolysis, heterolysis and photoionization [11]. Photosensitized photodegradation is based on the sorption of light by a molecule, thereby it transfers energy from its excited state to the pesticides. Photosensitization may also involve redox processes, such as the photo-Fenton reaction, where there is electron or atom transfer to produce free radicals [11].

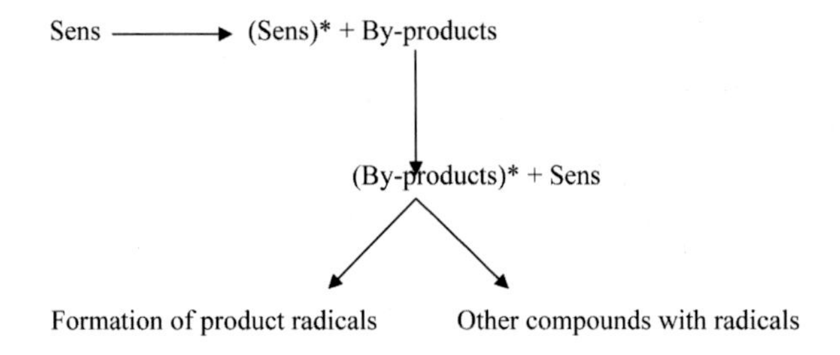

Another process known as photocatalytic degradation is a well-known process that involves cyclic photoprocesses in which the pesticides photodegrade, but spontaneous regeneration of the catalyst occurs to allow the sequence to continue indefinitely until all the substrate is destroyed [11].

6.4.1 Mechanism of degradation by hydroxyl radicals

There are different routes that hydroxyl radicals follow to degrade organic compounds or pesticides [11].

1. Addition of hydrogen peroxide that undergoes homolysis

$$H_2O_2 + h\nu \rightarrow 2HO^{\cdot\cdot}$$

2. Photolysis of ozone may generate atoms of singlet oxygen that further react with water to generate HO

$$O_3 + h\nu \rightarrow O_2 + O^{\cdot\cdot}$$
$$O^{\cdot\cdot} + H_2O \rightarrow 2HO^{\cdot\cdot}$$

3. Radiolysis of water

$$H_2O \rightarrow H^+ + HO^{\cdot\cdot} + \text{radicals}$$

4. Aqueous photolysis of Fe^{3+}, generated by oxidation of Fe^{2+} by H_2O_2

$$H_2O_2 + Fe^{2+} \rightarrow Fe^{3+} + OH^- + HO^{\cdot\cdot}$$

$$Fe^{3+} + H_2O + h\nu \rightarrow Fe^{2+} + HO^{\cdot\cdot} + H^+$$

6.4.2 Degradation of some pesticides in sea water

The photodegradation of the pesticides like aldicarb, carbofuran and carbaryl in distilled, pond and artificial seawater samples with or without humic acids was studied [12]. Photosensitizing effects occurred with carbaryl and carbofuran using different types of water, for example, pond water, seawater and seawater containing humic acids, which indicates that under real environmental conditions photodegradation is enhanced. Vebrosky et al. [13] studied the photodegradation of dicloran in artificial seawater and stated that the half-life is 7.5 h. No external factors affect photodegradation in freshwater with little variation in the formation of by-products. There are a few factors that affect the degradation of pesticides in seawater like irradiation condition, pH, media, presence of inorganic ions, photosensitizers and other subsidiary particles.

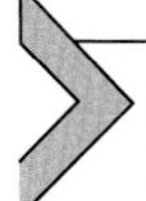

6.5 Nanocomposite membranes in removing pesticides from water

6.5.1 Introduction to nanocomposite membrane in water purification

In membrane technology pores in the fabricated membrane play a vital role in removing target pollutants from water. Numerous forms of membranes with diverse pore sizes are engaged in water treatment process, for example, microfiltration, ultrafiltration, reverse osmosis and nanofiltration membranes. There are some issues to be addressed in the existing membrane purifications like selectivity, permeability and low resistance to fouling. To address this issue a new type of membrane was fabricated called nanocomposite membrane (NCM). NCM is an innovative type of membrane prepared by merging complex constituents with nanomaterials. According to the membrane assembly and position of nanomaterials, NCM are classified into four groups: (1) conventional nanocomposite, (2) thin-film nanocomposite (TFN), (3) thin-film composite (TFC) with nanocomposite substrate and (4) surface-located nanocomposite. Exact water purification can be achieved by calibrating the chemical characters and distinct functionalities in them. There are two main factors that define the membrane separation process like pore size and porosity. NCM is a mixture of material that can have nanoscale inorganic and/or organic solid phases in a porous structure. Nanomaterials can improve numerous characteristics like mechanical strength, thermal stability, antifouling properties, permeability and selectivity, which are added advantages for the membrane separations process. Various constituents such as carbon nano tubes (CNTs), graphene and grapheneoxide (GO), silica and zeolites, metal and metal oxides, polymers, dendrimers and biological nanomaterials are used in NCM to improve water purification performance [14].

6.5.2 Formation of nanocomposite membrane

There are various ways to form NCM. Some of them are:

- Blending
- Interfacial polymerization
- Phase inversion
- Surface grafting
- Air drying

Among these procedures interfacial polymerization is the most commonly employed and is highly efficient for cleaning water pollutants. Fig. 6.3 shows different procedures for preparing nano composite membranes.

6.5.3 Application of nanocomposite membrane in removing pesticides from water

Organophosphorous pesticide malathion has been removed using a copper–chitosan nanocomposite (biopolymeric waste from the marine industry). It was found that pores play a vital role in removing these chemicals from water. The addition of copper ions increased the sorption capacity. It was found that other pesticides like methyl parathion and parathion can be removed under normal conditions with respect to experimental agenda [15]. Also some research findings state that such NCMs can be used for extracting pesticides like organochlorine from seawater. It was found that

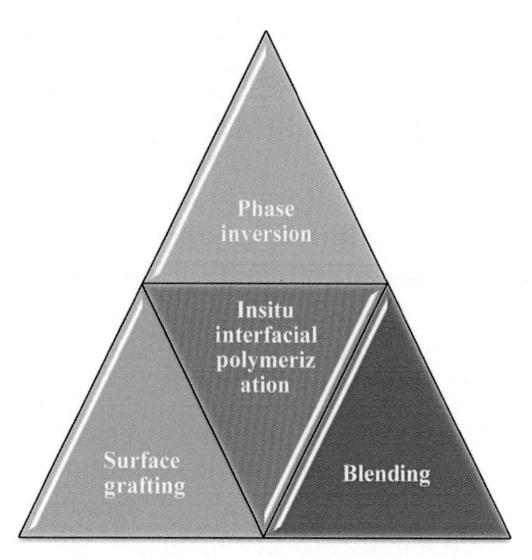

Figure 6.3 Methods for forming NCM. *NCM*, Nanocomposite membrane.

polyaniline/polypyrrole/cellulose nanocomposite filters can extract organochlorine from seawater. The limit of detections of heptachlor, aldrin, dieldrin, endrin and 4-DDT were found to be 0.39, 0.28, 0.47, 0.51 and 0.31 μg/L, respectively [16].

Also certain micropollutants have been removed from artificial seawater using an ultrafiltration membrane coupled with carbon nanotubes in order to prevent fouling. The provision of carbon nanotubes may increase the membrane flux, thereby promoting the removal of the target pollutant from water [17]. Coastal water from the northwest Mediterranean Sea showed potential risk of containing pollutants like organochlorinated pesticides, alkyl phenols and phthalates. It was found that such organic compounds pose a serious risk to fish, algae and the sensitive mysid shrimp-like organisms in seawater [18].

6.6 Phytoremediation—an advanced biological treatment

Phytoremediation is an innovative technology that has emerged as a promising tool which uses the ability of plants to remove pollutants in different matrices like soil, water and the atmosphere. It is a kind of bioremediation that uses vegetation as it tool in clearing pollutants from water. It is being promoted because it is economically effective in remediating a contaminated environment. There are several plants used for phytoremediation, for example, *Typha latifolia* (L.), *Leersia oryzoides* (L.) Sw. and *Sparganium americanum* Nutt, for the mitigation of atrazine, diazinon and permethrin in soil and the use of *Pseudokirchneriella subcapitata* and *Chlorococcum* sp. for the transformation of fenamiphos and its metabolites in waterways [7]. It is a cost-effective and eco-friendly technology that has the ability to remove environmental pollutants by using specific plant-associated bacterial strains from water resources. Aquatic plants such as *Eichhornia crassipes*, *Lemna minor* and *Elodea canadensis* have been used in water treatment. Due to their photosynthetic activity and growth rate, plants are easy to harvest and can have a high absorption rate of pollutants. Factors like the chemical and physical properties of pesticide compounds and environmental parameters play vital roles in defining the removal efficiency of aquatic plants. A high removal rate of pesticides occurs when the pesticide mobility rate is proportional to surface adsorption on plants [19].

6.6.1 Mechanism involved in phytoremediation

Generally there are four mechanism involved in removing organic pollutants from water through this technology. They are [19]:

- Direct uptake and accumulation of contaminants and subsequent metabolism in plant tissues.
- Transpiration of volatile organic hydrocarbons through the leaves.
- Release of exudates that stimulate microbial activity and biochemical transformations around the root system.
- Enhancement of mineralization at the root–soil interface that is attributed to mycorrhizal fungi and microbial concortia associated with the root surface [19].

6.6.2 Active plants in removing pesticides

6.6.2.1 Eichhornia crassipes

Known as water hyacinth. *E. crassipes* is a free-floating aquatic macrophyte that rapidly grows to a great depth. The common species has long pendant roots, rhizomes and stolon and it can grow to 40–100 cm in height. It has the ability to float due to the stems and leaves contain air-filled tissue that provide substantial buoyancy and it is also able to grow in severe conditions such as moist sediments for several months. This species was tested for the removal of malathion and it removed 10 ppm, around 56%.

From the test results it was observed that pollutant accumulated in different parts of plants (55–91% shoots and 74–81% in roots) started to degrade only after the start of growth of plants. This shows that *E. crassipes* takes up pesticides and its phytodegradation can be used as a potential economical and alternative biological method to remove pesticides from water bodies [19].

6.6.2.2 *Lemna minor*

Known as duck weed, *L. minor* is an aquatic macrophyte that withstands cold weather (1.7°C–35°C) and can grow rapidly within a week under optimum pH 6. The main advantage in using this plant is that it is cost-effective, has a long storage capacity, and requires minimal usage of chemical. It has the ability to remove dimethomorph until the concentration does not become too toxic and inhibits the depuration mechanism [19].

6.6.2.3 *Elodea canadensis*

E. canadensis is a free-floating, rapidly growing macrophyte aquatic plant. This species is commonly used in phytoremediation technology to treat organic pollutants like pesticides. It has been tested for the removal of three pesticides: flazasulfuron, dimethomorph and copper sulfate. The maximal removal rates for copper sulfate in *E. canadensis* was 16.5% within 3 days; dimethomorph was 5.5% within 2 days; and 12% within 4 days for dimethomorph [19].

6.6.3 Phytoremediation technology in organics

Clearing the environmental pollutants like pesticides using plants involves phytostabilization, rhizodegradation, rhizofiltration, phytodegradation and phytovolatilization [20].

6.6.3.1 *Phytostabilization*

Phytostabilization occurs in roots plants in which it exudates and combines with contaminants in the soil medium for degradation of pollutants followed by its availability in soil [20].

6.6.3.2 *Rhizodegradation*

This is the process to immobilize contaminants in groundwater and in soil by accumulation and absorption by roots, adsorption onto roots, or precipitation within the root zone. This procedure uses microorganisms to consume and digest organic substances for nutrition and energy [20].

6.6.3.3 Phytovolatilization

The uptake and transpiration of a contaminant by a plant, with release of the contaminant or a modified form of the contaminant to the atmosphere from the plant is called phytovolatilization [20].

6.6.4 Factors that influence uptake and chemical reactions pesticides by plants

Basic parameters like the physicochemical properties of the compounds, mode of application, soil type, climatic factors and plant species play crucial roles. Other factors are compounds that are absorbed through roots have to be translocated to the xylem which is done only by compounds that are slightly hydrophobic and whose log $K_{ow} > 3$. Microbial transformation of pollutants in the rhizosphere plays a major role in removing pollutants [21].

Three main reaction types are known to drive the transformation of pesticides in plants. They are [21]:

- Oxidative reaction
- Hydrolysis
- Conjugation
- Epoxide formation [21]

Biological oxidation of pesticides is catalyzed by a number of enzymes, such as cytochrome P450, peroxidase and flavin-dependent monooxygenase. These microsomal enzymes play a crucial role in biosynthesis and breakdown of a wide variety of endogenous and exogenous compounds through oxygenation or hydroxylation [21].

6.6.5 Enhancement of phytoremediation

A recent development in biodegradation is the use of genetically modified plants. Applying this idea in a phytoremediation approach that utilizes nonfood plants with enhanced ability to clean up pesticide-polluted environments may have a beneficial effect on the public perception of biotechnology. Enhancement can be done by the selective manipulation of a plant's growth and capacity to remove or degrade pesticides through enzymes or any other mode. To illustrate, a pesticide-degrading enzyme from another species can be engineered in plants. Certain insecticide-resistant insects have enzymes like esterases and cytochrome P450 that have the ability to kill or induce enhanced detoxification of insecticides. The main focus should be on selecting plants that can hyperaccumulate or biodegrade organic chemicals. It should possess an enhanced ability to absorb and bind a specific pesticide by incorporating a gene for an antipesticide antibody fragment. The single chain variable fragment (scFv) of immunoglobulins contains the combining site of an antibody toward a given analyte. Through incorporation of the relevant gene, this "subunit—antibody" can be expressed in plants to enhance their capability to bioaccumulate large quantities of the compound. Through appropriate design of immunogens, it may even be possible to engineer a group-specific antibody fragment that binds to a number of structurally similar chemicals [21]. There are lots of studies reported that use gene transfer in plants to induce detoxification of harmful pollutants by increasing the capacity to bind large quantities of specific pesticides from contaminated sites. Studies on the development of generic antibodies reactive to a number of pesticides have been reported. Research is developing plants to absorb multiple pesticides for degradation [21].

There are studies on transgenic plants expressing pesticides degrading enzymes, such as an extracellular fungal enzyme, the laccase of *Coriolus versicolor*, into tobacco plants. The transgenic plant was able to remove pentachlorophenol by producing laccase and secreting it into the rhizosphere. Likewise, the decontamination of organophosphorus pesticide using a bacterial organophosphorus hydrolase (OPH) gene in tobacco plants has been executed. The transgenic plants degraded more than 99% of methylparathion after 14 days of growth [19].

6.6.6 Limitation of phytoremediation

As discussed above phytoremediation offers lots of advantages like being low-cost, low maintenance, environment-friendly, and a renewable resource for remediation of a contaminated environment. There are also disadvantages as the technology would probably have little impact in situations where low levels of pesticides are widely distributed on the site. Even under favourable conditions plant growth and degradation of pollutants may not exceed a certain rate. These limitations require integration of phytoremediation with more immediate clean-up options, as well as more economical utilization of the biomass produced for nonfood purposes. There is also a need to find ways to enhance the absorption and degradation of pesticides in plants. As microbial activity in rhizosphere helps in the enhancement of uptake and transformation of pesticides in plants, a combination of microbial bioremediation and phytoremediation is likely to be more successful in the field [21].

6.7 Mycodegradation

Degradation of organic compounds using fungi is called mycodegradation. The process of degradation of pesticides by fungi may involve the formation of toxic or nontoxic by-products that purely depend on the ecosystem and its influencing parameters. To illustrate, *Aspergillus niger* showed high biodegradation of malathion pesticide. In another study *Aspergillus flavus* and *A. sydowii* were capable of degrading pirimiphos-methyl, pyrazophos and malathion, even at high concentrations (1000 ppm), utilizing these compounds as their sole phosphorus and carbon sources, and releasing the phosphorus moiety from these pesticides by means of their phosphatases. These reaction are carried out through enzymatic degradation [22].

This section deals with marine fungi and their ability to degrade pesticides. Generally marine organisms like fungi and bacteria are rich in novel enzymes. The enzymes derived from marine fungi are protein molecules with unique properties as they are derived from an organism whose natural habitat is saline or brackish water. Organisms from the marine environment are good source for proteases, carbohydrases and peroxidase. These enzymes can act as biocatalysts with properties like high salt tolerance, hyperthermostability, barophilicity and cold adaptability [22].

6.7.1 Fungal development

Development of fungi and its growth primarily depends on the variety of inorganic and organic compounds in the growing culture. Carbon is one of the most important elements for microbial growth, as carbon compounds provide energy for cell growth and serve as the basic units to build the cell materials. Nitrogen is also essential to the organisms, as well as other elements (hydrogen, oxygen and phosphorus). Pesticides like organophosphate have carbon, oxygen, sulfur and phosphorous in their structure that serve as a food for fungi. Some marine filamentous fungi are *A. sydowii* CBMAI 934, *A. sydowii* CBMAI 935, *A. sydowii* CBMAI 1241, *Penicillium decaturense* CBMAI 1234, *Penicillium raistrickii* CBMAI 931, *P. raistrickii* CBMAI 1235 and *Trichoderma* sp. CBMAI 932. These are multicellular organisms which grow as mycelia that consist of microscopic filaments called hyphae. The optimum pH that favours fungal growth and inhibits bacterial growth is 3.6–5.6 [22].

6.7.2 Mechanism of mycodegradation

The major mechanism involved in the degradation of organophosphorous compounds involves the induction of intracellular and extracellular enzymes. Fungi showed complete degradation of organophosphorous compounds of around 95%–99% at very high concentrations like 65 ppm. This shows that fungi are resistant to high concentrations of pesticides [22]. Fig. 6.4 depicts the mechanism of degradation of pesticides in fungi that have been isolated from the marine environment.

6.8 Conclusion

Pesticide exposure to fish and other aquatic life may be a more widespread problem than is reported and documented in many cases. Pesticides destroy the underwater conditions, leading to scarcity in young fish counts and the decolouring of many fishes. Though many treatment methodologies have been developed, the main progress lies in the prevention of the entry of pesticides into water. There are numerous acts that have been passed to control the release of such harmful contaminants into water. For example, US Congress passed the Endangered Species Act (ESA) in 1973 to protect animals and plants that are in danger of becoming extinct and to protect their habitat. The ESA requires that any action authorized by a

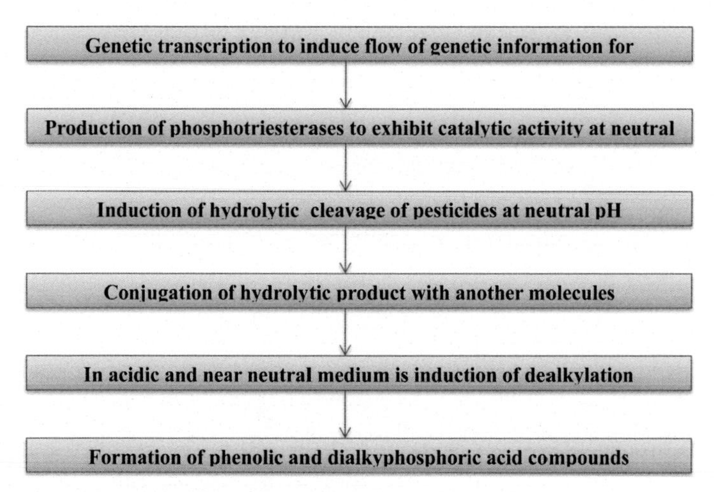

Figure 6.4 Mechanism of fungi degradation.

federal agency, such as the registration of pesticides, does not harm threatened or endangered species or their habitats. Environmental Protection Agency (EPA) plans to use limited pesticides in order to safeguard the habitats.

Pest management, such as integrated pest management (IPM), has been used to reduce pest populations. Factors such as groundwater contamination, the increasing cost of agricultural chemicals, consumer concerns about pesticide residues in and on foods, and concern for the environment encourage the use of IPM. Many fishes that survive in wetlands can be prevented from being harmed by pesticides by considering whether to use the pesticides or not. The best environmental management plans can be used to increase or improve the water quality. The use of microbial removal is highly recommended because it requires very much less cost compared to other removal methods. Also degradation will be complete as microbes use these chemicals as their food for survival and degrade them completely.

References

[1] Mensah KP, Palmer GC, Muller JW. Lethal and sublethal effects of pesticides on aquatic organisms: the case of a freshwater shrimp exposure to Roundup®. Pesticides 2014;. Available from: https://doi.org/10.5772/57166.

[2] Butler PA. Pesticides in the marine environment. J Appl Ecol 1966;3:253. Available from: https://doi.org/10.2307/2401464.

[3] Ernst W. Effects of pesticides and related organic compounds in the sea. Helgoländer Meeresuntersuchungen 1980;33(1-4):301–12. Available from: https://doi.org/10.1007/bf02414756.

[4] Lew S, Lew M, Biedunkiewicz A, Szarek J. Impact of pesticide contamination on aquatic microorganism populations in the Littoral zone. Arch Environ Contam Toxicol 2012;64(3):399–409. Available from: https://doi.org/10.1007/s00244-012-9852-6.

[5] Day KE. Pesticide residues in freshwater and marine zooplankton: a review. Environ Pollut 1990;67(3):205–22. Available from: https://doi.org/10.1016/0269-7491(90)90187-h.

[6] Satish GP, Ashokrao DM, Arun SK. Microbial degradation of pesticide: a review. Afr J Microbiol Res 2017;11(24):992–1012. Available from: https://doi.org/10.5897/ajmr2016.8402.

[7] Marican A, Durán-Lara EF. A review on pesticide removal through different processes. Environ Sci Pollut Res 2017;25(3):2051–64. Available from: https://doi.org/10.1007/s11356-017-0796-2.

[8] Cotham WE, Bidleman TF. Degradation of malathion, endosulfan, and fenvalerate in seawater and seawater/sediment microcosms. J Agric Food Chem 1989;37(3):824–8. Available from: https://doi.org/10.1021/jf00087a055.

[9] Vela N, Pérez-Lucas G, Fenoll J, Navarro S. Recent overview on the abatement of pesticide residues in water by photocatalytic treatment using TiO2. Appl Titanium Dioxide 2017;. Available from: https://doi.org/10.5772/intechopen.68802.

[10] Pandit G, Mohan Rao A, Jha S, Krishnamoorthy T, Kale S, Raghu K, et al. Monitoring of organochlorine pesticide residues in the Indian marine environment. Chemosphere 2001;44(2):301−5. Available from: https://doi.org/10.1016/s0045-6535(00)00179-x.
[11] Burrows HD, Canle LM, Santaballa JA, Steenken S. Reaction pathways and mechanisms of photodegradation of pesticides. J Photochem Photobiol B Biol 2002;67 (2):71−108. Available from: https://doi.org/10.1016/s1011-1344(02)00277-4.
[12] De Bertrand N, Barceló D. Photodegradation of the carbamate pesticides aldicarb, carbaryl and carbofuran in water. Anal Chim Acta 1991;254(1-2):235−44. Available from: https://doi.org/10.1016/0003-2670(91)90031-y.
[13] Vebrosky EN, Saranjampour P, Crosby DG, Armbrust KL. Photodegradation of dicloran in freshwater and seawater. J Agric Food Chem 2018;66(11):2654−9. Available from: https://doi.org/10.1021/acs.jafc.8b00211.
[14] Beyene HD, Ambaye TG. Application of sustainable nanocomposites for water purification process. Sustain Polymer Compos Nanocompos 2019;387−412. Available from: https://doi.org/10.1007/978-3-030-05399-4_14.
[15] Jaiswal M, Chauhan D, Sankararamakrishnan N. Copper chitosan nanocomposite: synthesis, characterization, and application in removal of organophosphorous pesticide from agricultural runoff. Environ Sci Pollut Res 2012;19(6):2055−62. Available from: https://doi.org/10.1007/s11356-011-0699-6.
[16] Mehdinia A. Preconcentration and determination of organochlorine pesticides in seawater samples using polyaniline/polypyrrole-cellulose nanocomposite-based solid phase extraction and gas chromatography-electron capture detection. J Braz Chem Soc 2014;25. Available from: https://doi.org/10.5935/0103-5053.20140190.
[17] Heo J, Joseph L, Yoon Y, Park Y-G, Her N, Sohn J, et al. Removal of micropollutants and NOM in carbon nanotube-UF membrane system from seawater. Water Sci Technol 2011;63(11):2737−44. Available from: https://doi.org/10.2166/wst.2011.602.
[18] Sánchez-Avila J, Tauler R, Lacorte S. Organic micropollutants in coastal waters from NW Mediterranean Sea: sources distribution and potential risk. Environ Int 2012;46:50−62. Available from: https://doi.org/10.1016/j.envint.2012.04.013.
[19] Chander PD, Fai CM, Kin CM. Removal of pesticides using aquatic plants in water resources: a review; 2018; IOP Conference Series: Earth and Environmental Science; 164.
[20] Tripathy S, Paul B, Khalua RK. Phytoremediation: proficient to prevent pesticide pollution. Int J Innov Sci Eng Technol 2014;1(10).
[21] Chaudhry Q, Schroder P, Reichhart DW, Grajek W, Marecik R. Prospects and limitations of phytoremediation for the removal of persistent pesticides in the environment. Environ Sci Pollut Res 2002;9(1):4−17.
[22] Da Silva NA, Garcia Birolli W, Regali Seleghim MH, Meleiro Porto AL. Biodegradation of the organophosphate pesticide profenofos by marine fungi. Applied Bioremediation - Active and Passive Approaches; 2013. doi:10.5772/56372.

CHAPTER SEVEN

Plastic litters removal

7.1 Introduction to plastic debris in marine water

Plastics are a diverse group of synthetic polymers which were discovered in the late 19th century. Their properties, like low density, durability, excellent barrier property and low cost, make plastics an ideal material in the field of manufacturing and packaging applications. Plastic's durability means it persists in the environment for many years. Its low density allows it to travel from its source elsewhere through wind or water. Hence it can be termed as an ubiquitous pollutant. Its potential impacts are witnessed in seawater and its organisms. Marine litters are solid particles that disposed of into the ocean. Plastic marine litters comprise artificial materials such as plastic bags, food packaging, cups, bottle and industrial items related to fisheries and aquaculture [1]. It is estimated that nearly 20 MT of plastics enter the ocean each year. Many anthropogenic activities are responsible for the release of plastics into the ocean. The accumulation rate of plastics in the ocean depends on many factors like the presence of large cities, shore use, hydrodynamics and maritime activities. This chapter covers the impacts due to plastics in the marine environment and some of the advanced technologies used in the removal of plastics from the ocean.

7.2 Sources of plastic litters

Generally sources of plastics entering the ocean may be either land-based or ocean-based. Land-based sources include mainly recreational use of the coast, general public litter, industry, harbours and unprotected landfills and dumps located near the coast, but also sewage overflows, accidental loss and extreme events. Ocean-based sources include commercial shipping, ferries and liners, both commercial and recreational fishing

Modern Treatment Strategies for Marine Pollution.
DOI: https://doi.org/10.1016/B978-0-12-822279-9.00005-1

vessels, military and research fleets, pleasure boats and offshore installations such as platforms, rigs and aquaculture sites. Factors such as ocean current patterns, climate and tides, the proximity to urban, industrial and recreational areas, shipping lanes and fishing grounds also influence the types and amounts of litter that are found in the open ocean or along beaches [1]. Fig. 7.1 represents sources of plastics litters that are entering ocean.

7.3 Different settlement of plastic litters in ocean components

7.3.1 Floating marine debris

This is the fraction of marine litter that floats due to its low density. A major part of floating debris is synthetic polymers whose floating nature depends on its physicochemical properties and environmental conditions. High-production-volume polymers such as polyethylene (PE) and polypropylene (PP) have lower densities than seawater, they float until they are washed ashore or sink because their density changes due to biofouling and leaching of additives [1].

7.3.2 Seafloor

Some plastic litters sink in water which cannot be assessed due to the depth of the water level. Deepsea surveys are important because nearly 50% of plastic litter items sink to the seafloor and even low-density polymers such as PE and propylene may lose buoyancy under the weight of fouling. The geographic distribution of debris on the ocean floor depends on parameters like hydrodynamics, environmental temperature geomorphology and human activities. It is found that the degradation of plastics

Figure 7.1 Sources of plastic litters.

in the deep sea is slow due to the absence of light, the low temperature and low oxygen concentration [1].

7.3.3 Microplastics

Particles that are less than 1 μm are termed as microplastics. Microplastics comprise a very heterogeneous assemblage of particles that vary in size, shape, colour, chemical composition, density and other characteristics. They are classified into primary and secondary microplastics. Primary microplastics are produced either for indirect use or as precursors. Secondary plastics result from the breakdown of larger plastic material into smaller fragment. Microplastics generally float at the sea surface because they are less dense than seawater. However, the buoyancy and specific gravity of plastics may change during their time at sea due to weathering and biofouling which varies due to their distribution across seawater [1].

7.4 Persistence of plastic litters in ocean

The persistence of plastics in water depends on buoyancy and its fillers. Plastics like PE and PPs are less dense than water, whereas polyvinyl chloride and PE terephthalate sink in seawater [1]. Generally plastics entering ocean undergo:

- Solar UV-induced photodegradation reactions
- Thermal reactions including thermooxidation
- Hydrolysis of the polymer
- Microbial biodegradation [1]

7.5 Some of the effects of plastics litter to marine biota

The most visible effect of plastic pollution on marine organisms concerns wildlife entanglement in marine debris and the ingestion of debris. Marine birds, turtles and mammals have received the most attention, but the consequences of entanglement and ingestion on other organism groups, for example, fish and invertebrates, are becoming more evident.

- Entanglement
- Smothering
- Ingestion of plastics
  - Intentional ingestion
  - Accidentally ingestion

Some of the effects in marine biota lead to direct morality like poor nutrition and dehydration. Also the plastics can be transferred up through the food chain which can have indirect impacts on other habitats and human health. Entanglement may lead to the death of some endangered species in the ocean. It has negative impacts on marine biota, thereby decreasing the populations. It also has effects on bigger animals, like sharks and seals [1].

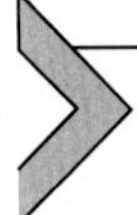

7.6 Common types of plastic litters found in marine environment

Plastics are a mixture of polymers and additives that increase the self-weight of plastics. Hence polymers that are commonly found in the ocean are PE, PP and polystyrene (PS). On average 92.2% and 95.8% of these polymers are found in seawater. The types of polymers that are found in water include PE terephthalate, polyvinyl chloride, polymethyl methacrylate, polyacrylonitrile and polyvinyl alcohol (PVA). Polymer composition is more diverse in seawater. Hence nearly 16.9% and 17.6% of polyamide and polyesters are seen in the ocean. Some fragments of products are seen in the ocean. 4.8% were lines, nets and ropes and 0.4% were nonfibre fishery and aquaculture items. 1.0% are packaging items, and a few items come from households and building and construction [2]. Fig. 7.2 shows the different types of plastics debris.

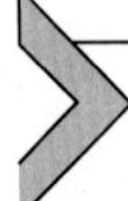

7.7 Biodegradation of plastic litters

7.7.1 Introduction to biodegradability of plastics

Biodegradation is defined as the action of microbes on any complex organic compounds that are found in any environmental matrix.

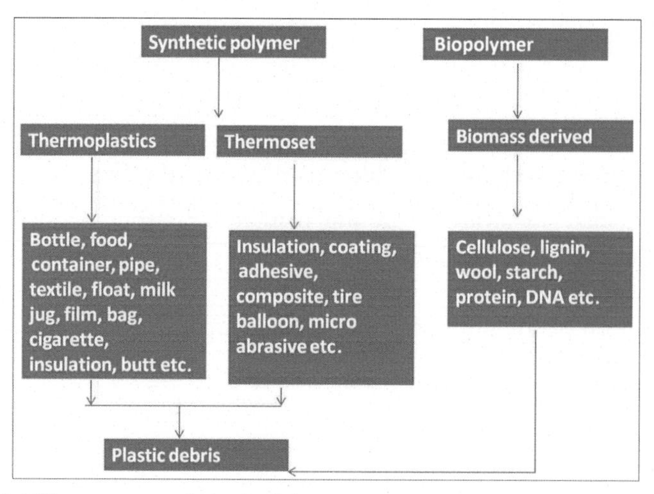

Figure 7.2 Different types of plastic debris.

Generally it comprises two states: primary biodegradation and ultimate biodegradation. The former is the biochemical transformation of compounds using microorganisms, whereas the latter is the degradation of material resulting in the complete mineralization or generation of biomass which is then applicable for other applications. Under biotic conditions mineralization will yield carbon dioxide and water, whereas under abiotic conditions it will yield methane and carbon monoxide. Major environmental factors that affect the biodegradability of plastics are physical–chemical factors, such as pH, temperature, soil structure, redox potential, moisture, nutrient and oxygen availability and the presence of inhibitors. Microbiological factors potentially affecting the biodegradability of plastics are distribution, abundance, diversity, activity and adaptation of microbiota [3].

7.7.2 Potentially biodegradable plastics materials

Generally plastics like PE and PP are easily biodegradable. Biodegradable plastics are synthesized using certain factors like using polymer composites, and introduce weak links into recalcitrant polymers to make them easier to hydrolyze. They use natural polymers like cellulose, using prooxidants to enhance biodegradation [3].

Polymers with weak links and bonds are susceptible to hydrolysis, photolysis, or oxidation. Amide and ester bonds are the weak links in polymers. These weak links may occur in the structure of the molecule, such as in polycaprolactone, polyamides and some polyurethanes.

Microbiologically produced plastics will have ester bonds that are susceptible to attack by microbial enzymes. The use of prooxidants like metals and lipids in polyolefins may enhance the degradation of plastics [3].

7.7.3 Measurement of biodegradation in plastics

Procedures that are available in the market for testing biodegradability include the following methods [3]:

- Enzymatic digestion
- Growth on plastic substrate as a carbon source
- Effect of soil burial
- Radiolabelled carbon [3]

7.7.3.1 Enzymatic digestion

The polymers are exposed to certain enzymes in vitro, such as hydrolytic enzymes, for breakage of bonds. This assay has a very limited ability to predict the environmental degradability of plastics as it relies on enzymes [3].

7.7.3.2 Growth on plastic substrate as carbon source

Plastic as a ground medium along with agar is grown with cultures like bacteria and fungi. The grown cluster of colonies are tested for the breakage of plastics, which they use as a source of food. The disappearance of the polymer on exposure to microorganisms suggests a transformation of the substrate [3].

7.7.3.3 Radiolabelled carbon

Radiolabelled carbon can be used to test the biodegradation of plastics. This test is very difficult as it is tough to synthesize the radiolabelled plastics which depend on special chemical synthesis [3].

7.7.4 Overview on biodegradation of plastics and its mechanism

Rather than natural polymers, synthetic polymers are the most difficult plastics to degrade completely. Some of the synthetic polymers belong to the following groups like polyesters, polyhydroxybutyrate (PHB), polycaprolactone (PCL), polylactic acid (PLA), polyurethane (PUR), PVA, nylon and PE. The major or common mechanisms that takes place for biodegradation of the plastics are the enzymatic or chemical reactions that

are taking place in cleaving them. Polymers like PE are highly resistant to biodegradation due to their [4]:
- highly stable C – H and C – C bonds;
- high molecular weight;
- lack of readily oxidizable or hydrolysable carbonyl, amide and C = C groups;
- lack of chromophores that can act as catalysts for synergistic photo and biodegradation;
- highly hydrophobic nature; and
- low biodegradability [4].

There are various biotic and abiotic factors that help in the biodegradation of PE [4].

7.7.4.1 Biotic factors

- Organisms like bacteria and fungi help in biodegradation.
- Biosurfactants produced by microbes attach on the PE surface.
- Biofilm grows on PE.
- Molecular level changes in PE due to extracellular enzymes by microbes.
- Uptake and assimilation of shorter chain PE via cell walls using intracellular enzymes [4].

7.7.4.2 Abiotic factors

- Factors like sunlight and photooxidation induce degradation.
- Addition of carbonyl radicals due to photooxidation.
- Fragmentation of PE by mechanical stress, heat, temperature, light, etc.
- Increase in hydrophilicity of PE causing diffusion of water molecules in to PE.
- Diffusion of oxygen in PE [4]. Fig. 7.3 illustrate the enzymatic degradtion pathway for polyethylene polymers.

Nylon which is a synthetic polyamide polymer is being degraded using microbes like bacteria and fungi. The pathway used for degradation by bacteria is illustrated below. Bacterial biodegradation of nylon 66 is hydrolytic process involving an oxidation reaction. Similarly in fungi an extracellular enzyme, fungus peroxidase, was used for degrading nylon [4]. Fig. 7.4 shows degradation pathway for nylon.

7.7.5 Biodegradation of synthetic plastic foams

Some of the microbes used for degradation of synthetic plastics foams are listed in Table 7.1.

There are various factors that affect the degradation of these foams:

- Evolutionary history behind these foams
- Enzyme—substrate complex formation
- Branched and linear chemical structure
- Cellular structure
- Hydrophobicity
- Molecular weight of these foams
- Mass transfer between cell and substrate
- Size of sample waste foam used for experiment
- Nature of biofilm and biomat formation which interrupt mass transfer action [4].

7.8 Role of floaters in plastic removal

Floaters are devices fabricated for ocean clean-up that operate by removing plastics from ocean on a large scale. The use of this device has successfully cleaned nearly 90% of plastics waste that pollutes the ocean. The main concept in fabricating this device is that it can float with the

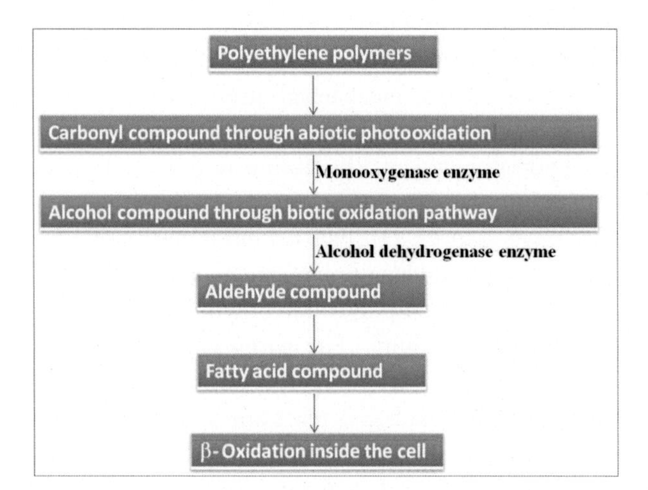

Figure 7.3 Synergistic degradation pathways for polyethylene polymers.

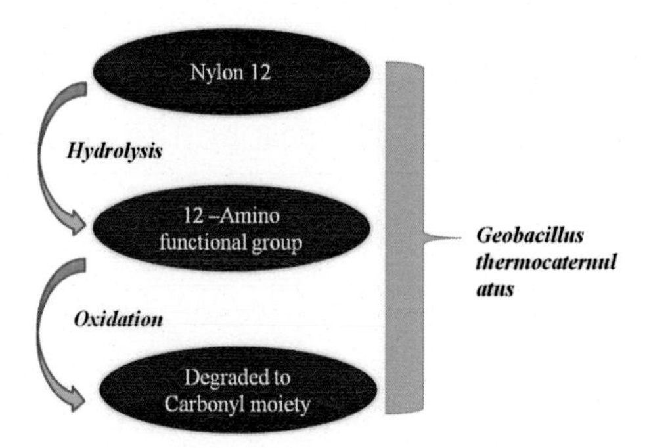

Figure 7.4 Biodegradation pathways for nylon using bacteria.

Table 7.1 Microbes degrading synthetic plastic foams.

Synthetic plastic foams	Nature of microbes	Microorganisms	References
Polyester PUR foams	Bacteria Fungi	*Pseudomonas aeruginosa* *Cladosporium resinae*	[4]
Polystyrene foams	Fungi	*Aspergillus niger, Aspergillus versicolor, Penicillium funiculosum, Chaetomium globosum, Aspergillus flavus*	[4]
Poly(butylene succinate) foams	Enzymatic	Buried in soil and performed enzyme process	[4]

ocean currents but at a slower rate than the plastics. This allows the debris to bump into the device's C-shaped arms and accumulate until a vessel arrives to take the plastics back to shore. The first device designed had trouble in retaining plastics in its arms and finally the end section of the device broke [5,6].

Later they launched a new design by using a parachute stating that it act as an anchor to slow down the device thereby allowing the system to retain plastics successfully until the vessels arrived to carry the debris back to land. The main drawback of this design was that plastics were being caught between the device and the screen which hangs in the water to collect the debris [5].

Finally they launched a device that allowed successful catching and retaining of plastics. The final upgrade in the device will be scaled up by increasing the size of the device to 600 m with a focus on long-term durability of the device as well as its ability to retain the plastics for a longer duration [5].

The ocean clean-up system is a U-shaped barrier with a net-like skirt that hangs below the surface of the water. It moves with the current and collects faster moving plastics as they float by. Fish and other animals will be able to swim beneath it [7].

7.9 Plastic catcher—for floating debris

A giant plastic catcher was launched for ocean clean-up to catch all the plastic debris and litter that floats on the water. But this model was broken due to ocean currents. Though it was a failed model, this is one of the innovations for cleaning plastics from ocean water. This giant floating pipe was 600 m long but failed to catch plastics. The model failed maybe due to the following reason [8]:

- A strong ocean current may demolish this giant catcher
- Device may imperil sea life
- Only a fraction of plastics are floating on the surface of the water [8]

7.10 In situ removal of plastics from marine system

There are three other major inventions that helped to clean ocean water. Among the three innovations, one was recently fabricated for ocean clean-up which is discussed in the previous section. Another invention is a bucket that sucks trash and oil out of the sea. This was created by two Australian surfers who named it the seabin. The seabin is a bucket with a pump and water filtration system that is designed to suck debris from the marine environment. The bucket includes an optional oil—water separator system that will pull oil right out of the ocean, then spit out cleaner water through the other side of the pump. They are installed in small numbers in certain places like San Diego and Finland [9].

The last invention is technology that turns plastic waste into oil. This technique uses a machine that works on pyrolysis which breaks down plastics into smaller molecules so that it can be turned into oil for further use. It is a kind of recycling and reuse method [9].

7.11 Marine debris in sea bed and its recovery

It is necessary to find out which types and the amount of plastics accumulate on the sea bed before recovering them from the contaminated site. The accumulation of plastics depends on the nature of the sand and its structural parameters. Hence it is first required to identify the presence of plastics which is done using probing equipment [10].

7.11.1 Identification of plastics using probing device

In shallow coastal water, exploration on the seafloor is done by two divers with a snorkel towed behind a small boat. For 100 m depths, a side scan sonar (SSS) was used together with a bottom trawl net, whereas from 100 to 2700 m manned submersibles were used [10]. Korea has developed the "Tow-Sled" for moderate depths between 500–1000 m. It is a kind of sleigh equipped with lights, cameras and long-range acoustic positioning systems, which provide information on the amount, type and location of marine debris. This system is attached to the boat with steel cables and moves by being towed on each side of the equipment base [10].

7.11.2 Removing plastics from shallow depth

The recovery of plastics depends on choosing the right equipment for removal from seabeds. It further depends on where the operation is going to take place, that is, is it a sandy or rocky place. Marine debris collection can be carried out by a crane located on a boat with interchangeable hooks, which may be curved to prevent debris slide or can be articulated to access narrow rocks. This method is operated in Korea around 130–150 days a year, and collects about 350 tons of marine debris annually [10]. The operation depends on the efficiency of machine and its operational parameters. It is highly influenced by coastal environmental parameters.

7.12 Plastic debris treatment

7.12.1 Plastic recycling—marine debris management

Plastic recycling is the process of recovering waste plastics and reprocessing the material into new useful products. Several factors greatly influence the recycling process, such as contaminants, molecular weight, mechanical properties and colour/transparency of the material. Different forms of recycling exist, such as mechanical, chemical and energy recovery recycling. Mechanical recycling is the best option for plastic waste treatment when the waste is good enough to make other good quality products from it. This type of process has lots of advantages: raw natural material is saved; energy and economic resources are saved; and greenhouse gas emissions from the manufacturing of plastics are reduced [11].

Once plastics are removed from marine water they undergo various processes such as incineration and fuel production. Since incineration involves the production of toxic gases, other technology developing fuel production and recycling are used. Before applying these technologies they are pretreated as the plastics contain salts and other contaminants. This process includes sorting, cutting, separating lead, grinding and cleaning of salts and sludge. This pretreatment process ensures mechanical stability, reduces the sodium salts and improves the quality of the material. The use of cleaning processes depends on the main process for the treatment of waste [10].

7.12.2 Mechanical recycling

It is carried out by various mechanical processes and the polymer structure of the product remains unchanged. The process consists of waste transformation by means of extrusion where plastics are melted and regranulated. This technique is widely applied because of its technical and economic feasibility. It has two different configurations: open loop and closed loop. In closed loop the quality and properties of recycled material remain close to the original material. Hence it can be used as product for high added-value product manufacturing. The input should be a single type of waste and slightly contaminated. In open loop, the input should be a single type of polymer material or a mixture of compatible plastics [11]. The section below discusses one mechanical mode of recycling of plastics debris, which is finally granulated and either disposed of or used as a product for other processes.

7.12.2.1 Fuel production using marine debris

Marine debris has high calorific value (4000–6700 kcal/kg). This feature makes it suitable for use as fuel. The analysis of the physical properties of refused-derived fuel (RDF) shows that marine debris contains a very high percentage of carbon and hydrogen, showing C: 73.58%, H: 6.304%, N: 0.338%, S: 0.391%, others: 19.387%. The process fuel production mainly involves grinding, water cleaning, dryer, silo and injection moulding extrusion. The residue from RDF can be used as feed in the recycling process [10].

The steps involved in the preparation of RDF are very common. It involves four stages [10]:

- preshredding;
- separation from preshredding;
- granulation line;
- separation using magnet and filtration; and thus
- final output [10]

7.12.3 Energy recovery

7.12.3.1 Thermal decomposition of plastic debris

Some of the marine debris cannot be recycled or reused. In such cases controlled and uncontrolled temperature treatment of plastic debris can be used. Processes like pyrolysis and combustion are considered to be thermal decomposition technologies that provide a reduction in volume of waste and also involve profitable energetic and/or chemical products. In incineration, cement kilns are categorized as controlled temperature technology, whereas open fire burning is uncontrolled temperature process. The substances emitted during uncontrolled plastic thermal degradation may create a serious hazard for human health and for the environment. In incineration, air emissions should be considered as it involves the formation of harmful gases due to incomplete combustion. Additionally the provision of a cleaning system is necessary in incineration plants to remove salts from marine debris. Since plastic debris from marine sources are landfilled currently, it is considered as a waste product from seawater with low calorific value. Nowadays this waste has great value and it is suitable for recycling, mainly by chemical or energy recovery [10]. Waste incineration is carried out for electricity production and district heating with efficiency above 90%. Plastic products are a good source of energy material due to their high calorific value. After incineration, the volume of waste can be reduced by 90%–99%, which is a big advantage when there is a shortage of space and landfilling is limited [11].

7.12.4 Chemical recycling

This is the process of breaking down the polymer structure to obtain the original monomer. Recycled polymer can be used as a stock material for new polymer production. Some of the chemical recycling process is depolymerization, partial oxidation and cracking. Depolymerization involves methanolysis, glycolysis and hydrolysis. The only drawback is the cost of separation of by-products, which is very high [11].

7.13 Biodegradation of marine plastic debris by plastisphere

This is the process of colonizing microorganisms over the plastic surface. Growth of microbial life on a plastic surface is called the plastisphere. Plastic debris is generally composed of PE at the ocean's surface, followed by PP and PS. Many organisms like bacteria, archaea, fungi and microbial eukaryotes are found growing on the surface of floating plastics. They are found only on certain plastics, like PE and PP, with high surface to volume ratios, such as rigid plastics, bundle fishing nets and ropes that are capable of floating on the surface of seawater, whereas other buoyant plastics like films sink underwater. It is observed that if photoautotrophic bacteria such as the cyanobacteria of the genera *Phormidium* and *Rivularia* dominate the subsurface plastisphere communities, the core microbiome of the seafloor and subsurface plastisphere seems to share taxa like Bacteroidetes (*Flavobacteriaceae*) and Proteobacteria (*Rhodobacteraceae* and *Alcanivoracaceae*) [12].

In laboratory studies the new plastics are incubated in marine conditions. Generally in the sea, plastics are rapidly covered by the "conditioning film" made of inorganic and organic matter, which is then rapidly colonized by bacteria (mainly *Gammaproteobacteria* and *Alphaproteobacteria*). There are various factors that affect the growth of biofilms over the surface of plastics like hydrophobicity and substratum properties like crystallinity and crystal structure, roughness, glass transition temperature, melting temperature and modulus of elasticity that assist in the selection of the bacterial community during the colonization stage. There is a unique difference in biodiversity and activity of bacterial community between biodegradable and nonbiodegradable plastics. It takes a long time for successful growth of bacteria or any microorganisms over the surface of plastics [12].

Factors driving the plastisphere composition and activities are complex, but are influenced by polymer type, surface properties and its size along with marine environmental conditions. Biodegradation is considered to be initiated after physical and chemical degradation that weakens the structure of polymers like roughness, cracks and molecular change followed by microbial activity. Two major process that are discussed in the biodegradation of plastics using microbes are biodeterioration and biofragmentation. Biodeterioration relates to the biofilm growing on the surface and inside the plastic, which increases the pore size and provokes cracks that weaken the physical properties of the plastic (physical deterioration) or releases acid compounds that modify the pH inside the pores and results in changes in the microstructure of the plastic matrix (chemical deterioration). Bio-fragmentation corresponds to the action of extracellular enzymes released by bacteria colonizing the polymer surface. These enzymes will reduce the molecular weight of polymers and release oligomers and then monomers that can be assimilated by cells [12].

7.14 Conclusion

This chapter deals with lots of recent innovations that are used for cleaning up ocean plastics. It is highly recommended to reduce the generation of waste at source level itself, thereby reducing the requirement to clean the ocean water. This can be achieved by improving waste management infrastructure and availability, stopping large items of plastic waste from entering the ocean, preventing microfibers from clothing and small plastic fragments and beads from entering wastewater by putting filters on washing machines, and finally a complete ban on the use of plastics that are not biodegradable.

References

[1] Bergmann M, Gutow L, Klages M. (Eds.). Marine Anthropogenic Litter; 2015. doi:10.1007/978-3-319-16510-3.

[2] Schwarz AE, Ligthart TN, Boukris E, van Harmelen T. Sources, transport, and accumulation of different types of plastic litter in aquatic environments: a review study. Marine Pollut Bull 2019;143:92–100. Available from: https://doi.org/10.1016/j.marpolbul.2019.04.029.

[3] Palmisano AC, Pettigrew CA. Biodegradability of plastics. BioScience 1992;42(9):680–5.

[4] Gautam R, Bassi AS, Yanful EK. A review of biodegradation of synthetic plastic and foams. Appl Biochem Biotechnol 2007;141(1):85–108. Available from: https://doi.org/10.1007/s12010-007-9212-6.

[5] <https://news.mongabay.com/2019/10/the-ocean-cleanup-successfully-collects-ocean-plastic-aims-to-scale-design/>

[6] <https://theoceancleanup.com/rivers/>
[7] <https://edition.cnn.com/2019/10/02/tech/ocean-cleanup-catching-plastic-scn-trnd/index.html>.
[8] <https://www.wired.com/story/ocean-cleanups-plastic-catcher/>.
[9] <https://www.huffingtonpost.in/entry/inventions-that-clean-the-ocean_n_5938-be94e4b0b13f2c66ee01?ri18n = true>.
[10] Iñiguez ME, Conesa JA, Fullana A. Marine debris occurrence and treatment: a review. Renewable and Sustainable Energy Reviews 2016;64:394–402. Available from: https://doi.org/10.1016/j.rser.2016.06.031.
[11] Horodytska O, Valdés FJ, Fullana A. Plastic flexible films waste management – a state of art review. Waste Manage 2018;77:413–25. Available from: https://doi.org/10.1016/j.wasman.2018.04.023.
[12] Jacquin J, Cheng J, Odobel C, Pandin C, Conan P, Pujo-Pay M, et al. Microbial ecotoxicology of marine plastic debris: a review on colonization and biodegradation by the "plastisphere". Front Microbiol 2019;10:865. Available from: https://doi.org/10.3389/fmicb.2019.00865.

Further reading

Balakrishnan and Sreekala, 2016 Balakrishnan P, Sreekala MS. Recycling of plastics. Recycl Polymers 2016;115–39. Available from: https://doi.org/10.1002/9783527689002.ch4.

CHAPTER EIGHT

Microplastics and its removal strategies from marine water

8.1 Introduction

Plastics are synthetic organic polymers that are derived from the polymerization of monomers extracted from oil or gas. Plastics are identified as a major pollutant in the marine environment and it is estimated that the amount of plastic debris entering the ocean is around 10% of total plastics produced globally. Large plastics debris is known as macroplastics. These macroplastics have serious effects on the marine environment and cause entanglement and damage to shipping equipment [1]. Some of the environmental impacts are the injury and death of marine birds, mammals, fish and reptiles resulting from plastic entanglement and ingestion, the transport of nonnative marine species (e.g. bryozoans) to new habitats on floating plastic debris and smothering of the seabed [1]. In recent years there has been increasing environmental concern on the effects and impacts of microplastics. Microplastics are small plastic fragments derived from the breakdown of macroplastics. Owning to their small size, they are considered to be bioavailable to organisms throughout the food web. Their composition and large surface area make them prone to adsorbing waterborne organic pollutants and to the leaching of plasticizers that are considered toxic [1]. This chapter covers the different types of microplastics that are seen in the marine environment, their environmental impacts and some of the removal strategies.

8.2 Microplastics and its categories

Microplastics are considered as fragmented pieces of large plastics that vary in size. They have numerous size ranges with diameters of <10 mm, <5 mm, <2 mm and <1 mm. They are released into environment through

Modern Treatment Strategies for Marine Pollution.
DOI: https://doi.org/10.1016/B978-0-12-822279-9.00003-8

two sources; primary and secondary microplastics. Primary microplastics are manufactured in the microscopic range. They are used in facial cleansers, cosmetics, air-blasting technology and as vector for drugs in medicine. They are marketed as microbeads or microexfoliates that vary in size, shape and composition. They are plastic pellets, beads, nurdles, fibres and powders used as industrial materials, personal care and cleaning products additives [1,2]. Secondary plastics are tiny plastics fragments derived from the breakdown of larger plastic debris both at sea and on land. Physical, chemical and biological processes can reduce the structural integrity of plastic debris, resulting in their fragmentation to form secondary plastics. Generally they are derived from the degradation of macroplastics under a weathering and aging process [1,2].

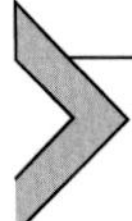

8.3 Sources and modes of transfer of microplastics in marine environment

Marine litter results from indiscriminate disposal of waste and the transfer of litter to the sea or ocean. The litter can be transferred directly or indirectly in to ocean. This plastics litter may be microplastics or macroplastics. Microplastics are used in cosmetics and air-blasting media and enter water through domestic and industrial drainage systems. Air-blasting technology involves blasting acrylic, melamine, or polyester microplastics scrubbers at machinery, engines and boat hulls to remove rust and paints. Generally, macroplastics will be trapped in wastewater treatment while microplastics will pass through the treatment system. The high unidirectional flow of the freshwater system drives the movement of plastics into the sea [1]. Coastal tourism, recreational and commercial fishing, marine vessels and marine industries (e.g. aquaculture and oil-rigs) are all sources of plastic that directly enter the marine environment, posing a risk to biota both as macroplastics and as secondary microplastics. Tourism and recreational activities account for the string of plastics along beaches and coastal resorts. Fishing gear is one of the most commonly noted plastic debris items with a marine source. Discarded fishing gear includes plastic monofilament line and nylon netting that are neutrally buoyant and can drift to varying depths in ocean. Further it causes entanglement of marine biota, known as ghost fishing [1]. Another notable source of plastic debris stems from the manufacture of plastic products that use granules and small resin pellets, known as 'nibs', as their raw material. Accidental spillage during transport, both on land and at sea, inappropriate use as packing materials and direct outflow from processing plants are some routes through which these raw materials can enter aquatic ecosystems [1].

Prolonged exposure of plastics to sunlight results in photodegradation; exposure to UV radiation results in the oxidation of the polymer matrix, leading to bond cleavage. High oxygen availability and direct exposure to sunlight will lead to rapid degradation, in time turning brittle, forming cracks and 'yellowing'. But the cold and haline conditions in the marine environment prohibit this photooxidation process. Upon continuous exposure these plastics lose their structural integrity, resulting in fragmentation from abrasion, wave action and turbulence. These fragments become smaller and smaller resulting in microplastics. It is considered that microplastics might further degrade to nanoplastics in size, although the smallest microparticle reportedly detected in the oceans at present is 1.6 μm in diameter [1].

8.4 Properties and distribution of microplastics

The major components of microplastics in the marine environment are polypropylene (PP), polyethylene (PE), polystyrene (PS), polyvinylchloride (PVC), polycarbonate (PC), polyamides (PA), polyester (PES) and polyethylene terephthalate (PET). The main properties that define the structural integrity of microplastics are the degree of crystallinity, density and glass transition temperature. The degree of crystallinity refers to the proportion of crystalline regions in polymers where the polymer chains are aligned with each other. They directly influence the mechanical property of polymers. Polymers can be separated into two categories, namely amorphous and semicrystalline. Semicrystalline polymers are characterized with high strength and high fatigue resistance. Amorphous polymers are soft and flexible and exhibit poor strength and poor fatigue resistance. Generally polymers change their physical form from glassy to rubbery state when the temperature rises above the glass transition temperature. The density of plastic defines the distribution pattern of microplastics in the water column. PE and PP are floating microplastics, as their densities are lower than water. PVC, PS, PET and PA are denser than water, so they tend to sink in the water column [2].

Microplastics are widely detected in inland lakes, estuaries, oceans and even remote areas, such as the Arctic Central Basin. Smaller microplastics

(0.02−1 mm) are more abundant than larger ones (1−5 mm) [2]. The occurrence of microplastics is abundant in the ocean. They are assessed through sampling techniques that are developed to detect the presence of smaller plastics debris. These techniques include:(1) beach combing; (2) sediment sampling; (3) marine trawls; (4) marine observational surveys; and (5) biological sampling [1]. Beach combing involves the collection and identification of all litter items along a specific stretch of the coastline. This is the simplest technique that is carried out by researchers and environmental awareness groups. This technique is particularly useful for determining the presence of macroplastics and plastic resin pellets, but microplastics, especially those too small to be observed by the naked eye, are likely to go unnoticed using such a technique [1]. Sediment sampling is a process to identify the presence of microplastics in benthic materials from beaches, estuaries and seafloors. Microplastics within the water column can be collected by conducting a trawl along a transect. Marine observational surveys allow divers or observers on boats and in submersibles to record the size, type and location of visible plastic debris. While this technique is effective at detecting macroplastics over relatively large areas, microplastics will often go undetected. Biological sampling involves identifying plastics fragments consumed by marine biota. By dissecting marine animals the presence of plastics can be identified and quantified [1].

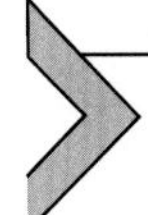

8.5 Changes in microplastics after degradation in marine ecosystem

Macroplastics and microplastics in marine water undergo various degradation processes through solar exposure, thermal aging, biofilm growth and oxidation. Degradation is defined as a series of chemical reactions that break the structure of plastic polymers. According to different weathering processes these degradations are termed as photodegradation, thermal degradation, biodegradation and thermooxidation degradation, respectively. This degradation process allows plastics to get fragmented and introduces secondary microplastics into marine water. During this process colour, surface morphological, crystallinity, particles size and density properties of microplastics are altered [2]. Fig. 8.1 gives some common ways for degrading plastics.

When plastic pellets are exposed to UV radiation they undergo photodegradation, during which yellow reaction products like quinone compounds are released. Hence the colour of microplastics shifts from white to yellow. However this process in the marine environment is severely retarded because of the lower oxygen concentration and temperature. Surface morphological changes in microplastics are due to fragmentation by mechanical forces and surface ablation by oxidative degradation. The mechanical forces such as hydraulic shear force and sand abrasion force could act on the surface of microplastics which consists of the amorphous component and the crystal component. The degradation rate of the amorphous component is faster than the crystal component. Thus cracks are generated in the surface and even in the deeper layers of the crystal component. Oxidative degradation takes place at the thin surface layer of microplastics as oxygen cannot diffuse to the inner layers [2]. The crystallinity of microplastics increases due to photo- and thermal oxidation. This is because (1) the amorphous component in polymers is more inclined to degrade than the crystalline component in the weathering/aging process; and (2) chemicrystallization, which progresses by random chain scission [2]. Fig. 8.2 depicts the properties that are changed after degradation of plastics.

8.6 Impacts of microplastics in marine environment

Microplastics are available in abundance in marine biota due to their small size making them prone to ingestion. The toxic effects of

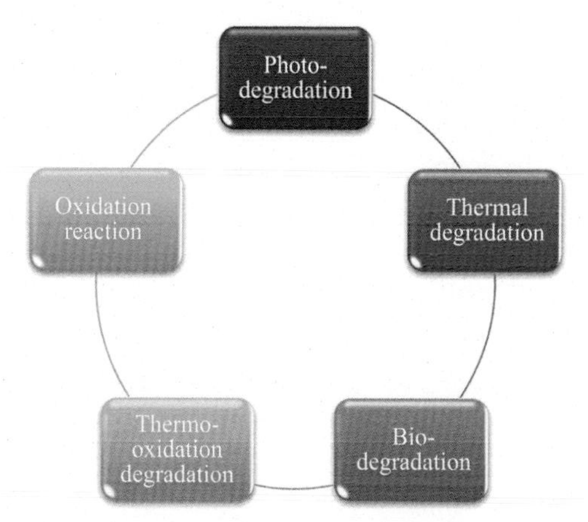

Figure 8.1 Degradation process in fragmenting plastics to microplastics.

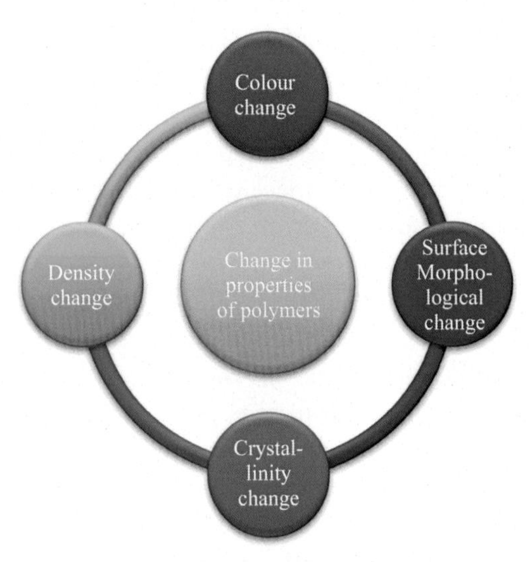

Figure 8.2 Changes in properties of plastic pellets after degradation.

microplastics are due to the contaminants leaching from the microplastics and the dissociation of extraneous pollutants that adhere microplastics.

8.6.1 Microplastics ingestion

Owing to their small size and presence in both pelagic and benthic ecosystems, microplastics have the potential to be ingested by a variety of marine biota; seabirds, crustaceans and fishes can ingest microplastics. Plastic fragments were first identified in the guts of sea birds in the 1960s, when global plastic production was less than 25 million tons per annum. The dissection of planktivorous mesopelagic fish, caught in the North Pacific central gyre, revealed microplastics in the guts. It is found that ingestion of nonpolluted microplastics have adverse health effects on biota like mortality and the reproductive process. Microplastics possess mechanical hazards to small animal, just like macroplastics through ingestion. Plastic fragments might block feeding appendages or hinder the passage of food through the intestinal tract or reduce the food uptake. Once microplastics are ingested, there is the potential for microplastics to be absorbed into the body through the digestive system via translocation [1]. They are ingested by marine organisms like zooplanktons, mussels, oysters, corals, fish, turtles and seabirds. Physically, microplastics will cause mechanical damages to the organism; microplastics could cause the blockage of the digestive tract, intestinal damage (including the cracking of villi and the splitting of enterocytes) and even alter the filtering activity and phagocytosis of organisms, finally leading to death of the organism. They could accumulate in food web by predation. Chemically they could accumulate and sorb pollutants. Biologically, microplastics are inclined to be colonized by microorganisms. Microplastics could influence the evolution of microbial communities and the exchange of genes between bacterial taxa [2].

8.6.2 Microplastics leachates

Plasticizers are additives that are added during the manufacture of plastics to change their properties and extend the life of plastics by providing resistance to heat, oxidative damage and microbial degradation. These additives are potentially hazardous to the environment as they leach harmful chemicals during prolonged exposure to sunlight. Incomplete polymerization during the formation of plastics allows additives to migrate away from the synthetic matrix of plastic. The degree to which these additives leach from plastics is dependent on the pore size of the polymer matrix, which varies by polymer, the size and properties of the additive and environmental conditions. Due to the large surface area to volume ratio of microplastics, marine biota may be directly exposed to leached additives after microplastics are ingested. Such additives and monomers may interfere with biologically important processes, resulting in endocrine disruption, which in turn can affect mobility, reproduction, development and carcinogenesis [1].

8.6.3 Microplastics with adhered pollutants

Microplastics with their large surface area to volume ratio, they can adhere to pollutants like heavy metals and persistent organic chemicals. Phenanthrene readily sorbs to small plastics, preferentially to polyethylene due to larger molecular cavities in this polymer. This is due to a good adsorption coefficient between plastics and phenanthrene. Microplastics coated with organic pollutants can be transported across the ocean polluting marine ecosystems. During ingestion they may release harmful toxins within the host body. Prolonged presence within the organism may magnify it and lead to carcinogenesis and other lethal effects [1].

8.6.4 Aggregation of biofilm covered microplastics with marine biogenic particles

Aggregate formation strongly depends on the coagulation efficiency of organisms (stickiness) and the particles involved. Microplastics are colonized by marine microorganisms, thereby forming biofilms. Biofilms are matrices of extracellular polymeric substances that increase the stickiness to external particles. Further aggregation with biogenic particles will increase the density, thereby increasing the sinking rate. This process makes aggregates available to deepsea segments, thereby inducing the redistribution of microplastics from the surface of the ocean to the deep sea [3].

8.7 Treatment protocols for marine microplastics

The presence of microplastics in the marine environment is more serious than other pollutants as they affect marine organisms and cause lethal effects in them like death, bioaccumulation and entanglement. They are categorized as persistent material because of their low degradability. The degradation of microplastics depends on their nature and chemical structure [4]. Hence their removal from environment is very important. Some of the removal methods are discussed below and include:

- sorption of microplastics on algae surface;
- using filtration technology like membrane process;
- chemical methods like coagulation and agglomeration process; and
- biological removal [4].

8.8 Sorption/bioflocculation of microplastics on algal surface and with biopolymer

Marine macroalgae are often mentioned as seafood that are rich in polysaccharides, minerals, vitamins, proteins and other compounds like fucoxanthin and fucoidan. A study investigated using seaweed as a potential sorbent for polystyrene microplastics [5]. The results revealed that a high sorption of microplastics of nearly 95% was found near the cut surfaces of the seaweed. This is because of the release of alginate compounds from the cut surface of the seaweed. Alginate, which is an anionic polysaccharide substance, was responsible for improving the adherence property of polystyrene particles on the seaweed surface. The effective sorption is due to the surface charge of microplastics and the surface characteristics of seaweeds [4].

The sorption of polystyrene particles onto seaweed's surface may be influenced by the particles' surface. To justify this, two studies were conducted, one using 20 nm size plastic particles on cellulose film and microalgae and the other using 20–500 nm particle on unicellular algae. Both studies concluded that neutrally charged and positively charged particles have a high binding affinity towards microalgae compared with the negatively charged group. This charged surface is due to the presence of functional group within the plastics. Hence it can be stated that the affinity of the algal surface to plastics will depend on specific properties and the surface charge of both plastics and algae [5]. The plausible physical chemistry for the adsorption, much in favour of the positively charged plastic, includes electrostatic interaction, hydrogen bonding and hydrophobic interaction between the algal species and the plastic [6]. Also there is a finding that the physical adsorption of plastic particles on the algal surface affects algal photosynthesis, because of the physical blockage of light and air by plastic particles. But this hindered photosynthetic algae can be used for the removal of toxic algal blooms from water [6].

Factors like hydrogen bonding, hydrophilicity, particle size and increasing specific surface area ratio affects the sorption behaviour. However, the presence of microplastics may increase biotoxicity, increase the dissolved organic matter in the environment and further influence marine environmental cycles like the carbon cycle [7]. Increased bioavailability may increase persistence and thereby be harmful to human life. This is because properties of microplastics vary by functional groups, specific area, density and crystallinity [7]. Hence all these parameters are considered when it comes to sorption using microplastics.

The role of microalgae with biopolymers like extracellular polysaccharide (EPS) is used in removing microplastics from water by employing it as composite flocculants. EPS are biodegradable displaying viscous gel-like structures in which the polymer molecules are assembled to form tangled networks or covalently cross-linked networks. They establish London dispersion forces, electrostatic interactions and hydrogen bonding in the adhesion and cohesion of suspended solids, making them potential bioflocculants. Bioflocculants are also regarded as safe and biodegradable, with less sludge generation and no secondary toxin production [8]. The study investigated the use of microalgae with EPS as excellent flocculants. The study confirmed that EPS produced by *Cyanothece* sp. exhibits high bioflocculant activity in low concentrations. Also, the EPS displayed very favourable characteristics for aggregation, as the aggregates consist of microalga, EPS and microplastics. It highlights the potential of the microalgal-based biopolymers to replace the hazardous synthetic flocculants used in treating water, while aggregating and flocculating microplastics and could be a multipurpose, compelling, biocompatible solution to microplastic pollution [8]. Fig. 8.3 illustrate some of the factors that are influencing sorption of plastics on marine algae.

8.9 Effective removal using metal based salt coagulation/ultrafiltration

Plastics that are often witnessed in the marine environment are broken into fragments due to the prolonged exposure to mechanical action, biodegradation and photooxidation. When the size of the particle is less than 5 mm then they are termed as microplastics. Their presence in environment induces some common problems like hindrance in light due to suspended or floating microplastics (As microplastics are low in density). Hence it is mandatory to remove it from water using various treatments like coagulation, filtration and biological removal [9]. Coagulation with ultrafiltration has been discussed in various research articles because it is a simple, cost-effective and economical treatment.

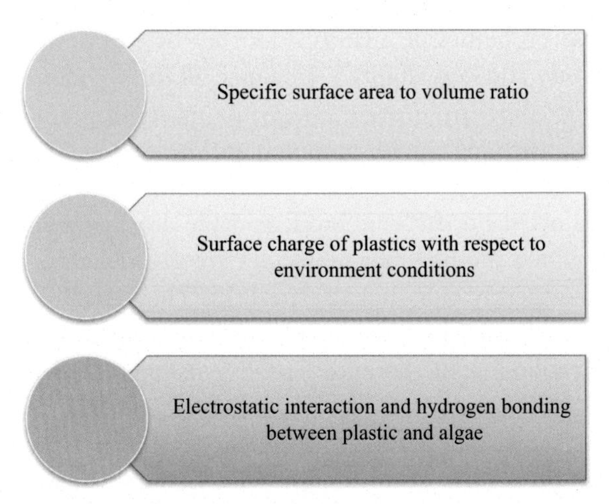

Figure 8.3 Factors influencing sorption of plastics on marine algae.

Coagulation is the process of forming flocs using external coagulating agents with microplastics to increase their specific weight for easy settling in water. Various coagulating agents like aluminium-based salts and iron-based salts are used in the coagulation process. Parameters like pH play important roles in floc characteristics because of the strong hydrolysis of coagulant. Easy removal of polyethylene microplastics have been achieved using Fe-based salts which is due to the dense floc formation and high adsorption ability because of the positively charged Fe-based flocs under neutral conditions. When combined with ultrafiltration complete removal was achieved. It is common to witness membrane fouling in the membrane filtration process because of the accumulation of plastics in the filtration zone due to particle sizes larger than the membrane pore size [9].

High dosages of aluminium- or iron-based salts showed less removal of microplastics and water conditions like turbidity and ionic strength barely influenced the removal rate. Similarly increasing dosage in the membrane process resulted in the aggregation of plastics leading to thick cake formation resulting in membrane fouling. This can be controlled since the larger the microplastic particles, the greater the roughness of the Al-based floc cake layer, leading to less severe membrane fouling. The microplastics removal behaviours exhibited during coagulation and ultrafiltration processes have potential application in treating water [10].

8.10 Electrocoagulation in extracting microbeads

Because of the small size of microplastics they are ingested by local marine flora which then face lots of health issues like improper digestion, malfunction of metabolic pathways and reduced morality. As they are the primary marine pollutant it is necessary to remove these from water at the source level itself. Cosmetic-derived microbeads can transfer adsorbed organic pollutants to the aquatic species that ingest them, thereby making cosmetic microbeads a serious pollutant [11]. Electrochemical techniques such as electrocoagulation, electroflotation and electrodecantation are a cheaper method that does not rely on microorganisms or chemicals. Electrocoagulation uses metal electrodes to produce coagulation electrically, making it simple and cheap. The main advantages of the electrocoagulation process are environmental compatibility, low capital costs, energy efficiency, sludge minimization, amenability to automation and cost-effectiveness [11].

The main principle behind electrocoagulation is producing metal ions from electrodes into water through electrolysis. These metal ions form coagulants that help in segregating microbeads from water. The most commonly used coagulants produced by EC are formed by the reaction of the metal ions (Fe^{2+} or Al^{3+}), with OH^- ions formed by electrolysis to produce metal hydroxide coagulants. These coagulants destabilize the surface charge of the suspended solids, breaking the colloids or emulsion that in turn come closer through van der Waals forces of attraction. Simultaneously coagulant particles form a sludge blanket that traps suspended solid particles. During the electrolysis process H_2 gas is liberated which helps in lifting the resultant sludge to the water surface [11]. Parameters like initial pH, current density and conductivity influence the operational conditions. Research findings show that the optimal pH for removal of polyethylene beads is 7.5, achieving around 99% removal. pH affects flocs formation which in turn affects time removal efficiency. It is found that charge neutralization is the main mechanism for the flocculation process [11].

The optimum conductivity was found to be 7.44–13.75 mS/cm that showed 99% removal in 40–60 min. Cost of electricity is the dominant factor in determining the operating cost of this process that further depends on the concentration of the electrolyte (NaCl) used for coagulant formation. The current density showed that it does not affect the efficiency of this process but at low current density it will improve the energy efficiency of electrocoagulation [11]. During operation microbeads undergo rapid charge neutralization during the first 15 min, which is able to quickly remove 50% – 80% of the initial load of microbeads within this time frame. In the remaining operating period flocculation mechanisms dominate and remove the remaining microbeads trapped in colloidal suspension. The combined effect offers low-cost treatment in removing microplastics from water. The most viable option for a large-scale industrial electrocoagulation cell for removing microbeads seems to be a two-stage, continuous electrocoagulation reactor/settler unit [11].

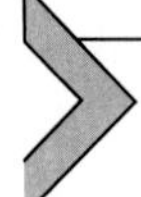

8.11 Alkoxy-silyl induced agglomeration—sustainable removal of microplastics

This method is a new technique in forming aggregates using polymers that are prepared using the sol–gel method. This induces easy separation as they can float readily on the surface of the water. This method is applied for various chemical preparations like organic–inorganic hybrid gels where there are many possibilities to steer the aggregation and condensation process by means of weak intermolecular forces (e.g. hydrogen bridge bonds, dipole–dipole interactions, etc.). In the case of template-like molecules, aggregated consistent structured hybrid silica gel is formed that is further functionalized, which can be used in medicine, sensors and optically active materials. At molecular level, the sol–gel process can be described for silicon alkoxides through hydrolysis and condensation. Hydrolysis as well as condensation can be represented as bimolecular nucleophilic substitution reactions. The mechanism differs according to the type of the catalyst used. Acids (H^+), bases (OH^-), or nucleophilic agents (F^-, N–methylimidazole, etc.) can serve as catalysts. Since the acidity of the protons in the remaining silanol groups is increased through the resulting siloxane formation, the base-catalyzed condensation mainly takes place between large, highly condensed species with SiOH groups and monomers, resulting in the formation of branched dense polymers [12].

Herbort et al. [12] state that bioinspired alkoxy-silyl, functionalized molecules act as adhesion reagents between the microplastic particles such as PE and PP. Through a sol–gel process, an agglomerate is formed which corresponds to 666 times the volume of the original particle and can be applied for removing it from water by increasing its density and settling period. This is a cost-effective separation method that can be applied in removing microplastics from water. This result is purely based on a characteristic molecular structure involving molecular organization in the precursor and hybrid silica gel. The agglomerates can be easily separated by a sand trap. This is a new technology that is completely independent of the type, size and amount of the trace substance concentration as well as of the external influences (pH value, temperature, pressure).

8.12 Membrane process for microplastics removal

According to the National Oceanic and Atmospheric Administration (NOAA), microplastics are particles of synthetic polymers of less than 5 mm in size that resist the natural degradation process. On the other hand, nanoplastics are particles with smaller dimensions, between 1 and 100 nm in size. The more common plastic materials found in water are PP, PE, PS, PVC, PC, PA, PES and PET; these are reversible thermoplastic polymers, highly recyclable materials that can be heated, cooled and shaped repeatedly [13]. The removal of plastics depends on some parameters like shape, size and mass of plastics particles. Though membrane processes and their application in removing microplastics are limited they find a special place in treatment methodologies due to the high-class performance. Some of the factors that affect the performance of membrane processes in removing microplastics are membrane material, pore size, thickness and surface area. Membrane process parameters include flux, transmembrane pressure, polarization concentration, cake layer formation and fouling and specific energy consumption [13].

Cross membrane technology (CMT) is known for its success in removing pesticides, bacteria, suspended solids, colour from water and has been shown to be a practical solution in cleaning microplastics pollution from the environment. CMT works as a physical barrier against micro-plastics, where the particulates cannot pass through the membrane on the basis of particle size. It allows only crystal clear commercially sterile, particle-free water to be discharged. All microplastics particulates and other separated impurities are separately retained at low volume for further treatment. This technology is cost-effective in terms of its ability to reduce water usage, waste volumes, energy and disposal costs and recover chemicals [14].

Membrane technology has been tested with various options in laboratory, one of which is the membrane reactor. The membrane reactor (MBR) is an advanced application that involves a membrane process and its principle in it. Using its versatile properties huge amounts of microplastics have been rejected in the carrier side of the MBR system distributed in water column and water surface. The finding stated that adsorption was one effect that favoured the removal of microplastics from water. The membrane used in this study has a pore size of 0.1 μm, the size of PVC added was <5 μm. Thus the microplastics could be totally rejected in the system [15]. One of the major drawbacks is the evolution of membrane fouling due to the formation of a cake-like layer, resulting in a decrease in removal efficiency of microplastics from water. This can be controlled by cleaning the fouled membrane using various technologies, such as physical cleaning using water, ultrasonic cleaning, alkaline cleaning and acid cleaning. This helps in the removal of the organic matter concentrated over the membrane. Studies found that irreversible membrane fouling in the MBR system with micro-plastics contamination was slightly higher than the absence of the microplastics condition. This suggested that it was possible for some tiny microplastics to enter the pores of the ultrafiltration membrane and increase the original membrane resistance. This implies that the microplastics-contaminated polluted water treatment with membrane technology could reduce the membrane service time [15].

Hence the membrane bioreactor process can be applied to treat microplastics-contaminated water and reject almost all microplastics from water. Membrane fouling caused by small microplastics can be cleaned by physical cleaning and thus the resistance can be recovered. One of the drawbacks that had to be addressed is the irreversible membrane fouling caused by small size microplastics that are present in the water. Another issue that kills membrane application is membrane abrasion caused by fragmented particles on the surface of the membrane. This usually arises in a cross-flow system like reverse osmosis. Irregular shapes of microplastics can damage membrane-like degradation of pores of membrane due to the high pressure induced on them [16].

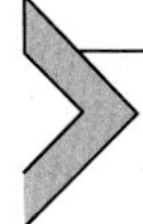

8.13 Function of density separation in extracting microplastics from water

Density separation is the first technique used to extract microplastics from water samples, and is applied to water contaminated with microplastics to limit its entry into treatment units. Commonly found microplastics are made of PE, PP, PVC and PES. It is possible to change the density of water to make microplastics float and remove them by skimming. The major principle is that the densities of plastics are very close to water, hence materials made of plastics are suspended in water making them complex to removal. Hence adding salts like NaCl and NaI will increase the density from 1.00 to 1.2 g/cm^3 and 1.8 g/cm^3, respectively, so that microplastics can float and thus can be removed more easily [16]. Since microplastics are smaller in size density separation is done in a static mode since flowing water will disperse these particles. The density of microplastics depends on their composition, not only the polymer they are made of but also any additives used during their manufacture or chemicals adsorbed on their surface. The difficulty is that the floating of microplastics purely depends on the density of water which is difficult is predict. Improved addition of salts to increase the density of water may be hazardous, such as NaI which is toxic for the environment.

Recovering the salts used during the separation steps and washing procedures for reuse is costly since the products are usually reprocessed until the desired purity is reached, this is one drawback for this process. This technology can be applied as a pretreatment in the water treatment process to prevent agglomeration in any filtration units like sand filtration or membrane separation [16].

8.14 Biological removal of microplastics

The process of fragmenting microplastics using microorganisms naturally is called biodegradation. It is a solution to increase the degradation rate with biological agents using plastic materials as source of carbon for growth. Plastic biodegradation is an ideal solution to remove microplastics from water, however environmental parameters such as water composition and temperature, plastic type and concentration and bacteria availability limit and slow down the process of biodegradation [16]. Microorganisms like picoeukaryotes, bacteria and archaea have the potential to biologically degrade microplastics in coastal sediments. A study by Padervand [4] in degrading high-density polyethylene secondary microplastics in seawater using two types of indigenous marine communities, *Agios* consortium and *Souda* consortium, gave many preliminary results. The density of plastics was found to be reduced by the *Souda* consortium. It is found that microplastics acted as a rich carbon source to feed the organisms, showing increased carbohydrate content in cell and decreased protein content.

Zalerion maritimum, a naturally occurring fungus in marine ecosystems, shows capacity for degrading PE microplastics based on the mass and size of microplastics. Degradation reaction kinetics by the fungus followed a fractional order that shows complex mechanisms are involved in fragmentation. Some of the bacteria extracted from coastal sediments, e.g. *Bacillus cereus* and *Bacillus gottheilii*, show good degradation for microplastics like polyethylene, polystyrene, polyethylene terephthalate and polypropylene. Degradation was measured with respect to weight loss of microplastics. The fastest degradation was observed at the rate of 0.0019/day in 363.19 days and the shortest degradation in 0.0016/day in 431.25 days [4].

Biodegradation is an effective option for eco-friendly degradation of plastics from water. Biodegradation involves a biological agent utilizing the organic polymer as a substrate for growth and energy, so that the end product of complete biodegradation will be microbial biomass. Kumar et al. [17] isolated microalgae like green algae, blue algae and diatoms for degrading PE found in water. The biological treatment of the selected microalgae on the low-density and high-density polyethylene sheets with their respective culture media showed that the microalgae proliferated more on the low-density than the high-density polyethylene sheets and the highest percentage of degradation was obtained from *Anabaena spiroides* treatment (8.18%). Thus photosynthetic microalgae which are most dominating microbes in the water system are found to be effective in colonizing the surface of PE microplastics and yield high degradation. It can be concluded that filamentous algae like blue–green algae are effective in degrading PE sheets that are commonly found in the environment.

Biodegradation of synthetic polymers often occurs in two environmental conditions, i.e. aerobic and anaerobic conditions. The extent of the degradation of polymers into CO_2, H_2O, N_2, H_2, CH_4, salts, minerals and biomass (mineralization) can be full or partial. Biodegradation is coupled to three essential criteria [18]:

- The environmental parameters, like temperature, pH, moisture and salinity must provide conditions that are necessary for biodegradation.
- The morphology of polymer particles must allow the attachment of microorganisms and the formation of biofilm provided structure of polymeric substrate should not hinder microbial actions.
- Microorganisms should be present to degrade the polymeric substance into proper monomers and finally mineralization [18].

The first step of biotic biodegradation is the cleavage of the polymer backbone and formation of smaller polymer units by extracellular enzymes. Secondly, smaller molecules are absorbed by microorganisms and metabolized to smaller products. In abiotic degradation, hydrolysis can result in intermediates that are then metabolized by microorganisms. Mostly aquatic plastics are very slowly degraded. Hence biodegradation is accompanied by a combination of biotic and abiotic degradation pathways [18].

8.15 Degradation using photochemical oxidation

Degradation of microplastics at the molecular level is initiated by UV radiation or by hydrolysis followed by chemical oxidation. The mechanism of photodegradation depends on the structural composition of polymers and their type. After reacting with photonic groups the molecular weight of the polymer is decreased and the reacted groups become available for microbial degradation. Degradation using photooxidation is faster than any other reaction [18]. In situ these reactions are carried out by exposure to photocatalysts that assist the reaction. Photocatalytic degradations occurs when a semiconductor is exposed to a source of light-emitting photons with equal or higher energy than the band gap, causing the generation of holes (h^+) and excited electrons. These holes are combined with water to form hydroxyl groups which are highly oxidizing species that are capable of degrading organic pollutants like polymers that are found in water [19]. The most commonly employed photocatalyst is TiO_2 because of its emerging capacities and unique properties that define the chemical behaviour. Green synthesis of protein-doped TiO_2 is used for degrading microplastics in a sustainable manner. The study used two semiconductors based on N-TiO_2 in which one presented an excellent capacity to promote degradation in aquatic and solid environment and the other presented a good capacity to promote the mass loss of extracted microplastics in an aqueous environment [19]. This process achieves complete mineralization of microplastics within a short span of time. The reaction time is less compared to other forms of degradation, but its efficiency lies in electricity consumption, material preparation protocols and mineralization products, as there are chances for the formation of toxic intermediates.

8.16 Miscellaneous treatment

8.16.1 Degradation using functionalized carbon nanosprings

Microplastics contamination in aquatic systems has emerged as a global issue due to its hazardous impact on the environment. More novel technologies have emerged for removing or degrading the pollutants. This is a combination of carbon catalytic oxidation and hydrothermal hydrolysis of microplastics over magnetic spring-like carbon nanotubes. This carbon hybrid showed perfect degradation of microplastics through catalytic activation of peroxymonosulfate to generate reactive radicals. This spiral structure and presence of graphite gave super stability to carbocatalysts in hydrothermal treatment. Hydrothermal treatment helps to withstand the temperature for a prolonged period of time in a closed environment. Hence this study showed the evolution of integrating the positive nature of carbon catalysis and nanotechnology to remove microplastics from aquatic zones [20].

8.17 Conclusion

Microplastics are formed by fragmenting plastics into the micrometer size through external factors. This is a new topic in the research field that is concentrated as a major global issue because of its serious hazardous impacts. The fragmenting process like thermal degradation, biodegradation, photodegradation and thermooxidative degradation assist in breaking polymer backbones into microplastics. Some of the removal technologies like sorption on microbes, coagulation, degradation using microorganisms, density separation, photooxidation and membrane filtration are discussed in terms of the removal of microplastics from water. All treatment processes depend on the structure of polymers, marine environmental parameters and the cost-effectiveness of the chosen treatment protocol.

References

[1] Cole M, Lindeque P, Halsband C, Galloway TS. Microplastics as contaminants in the marine environment: a review. Mar Pollut Bull 2011;62(12):2588−97. Available from: https://doi.org/10.1016/j.marpolbul.2011.09.025.

[2] Guo X, Wang J. The chemical behaviors of microplastics in marine environment: a review. Mar Pollut Bull 2019;142:1−14. Available from: https://doi.org/10.1016/j.marpolbul.2019.03.019.

[3] Michels J, Stippkugel A, Lenz M, Wirtz K, Engel A. Rapid aggregation of biofilm-covered microplastics with marine biogenic particles. Proc Royal Soc B 2018;285:20181203. Available from: https://doi.org/10.1098/rspb.2018.1203.

[4] Padervand M, Lichtfouse E, Robert D, Wang C. Removal of microplastics from the environment. A review. Environ Chem Lett 2020;3:807−28. Available from: https://doi.org/10.1007/s10311-020-00983-1.

[5] Sundbæk KB, Koch IDW, Villaro CG, Rasmussen NS, Holdt SL, Hartmann NB. Sorption of fluorescent polystyrene microplastic particles to edible seaweed *Fucus vesiculosus*. J Appl Phycol 2018;30(5). Available from: https://doi.org/10.1007/s10811-018-1472-8.

[6] Bhattacharya P, Lin S, Turner JP, Ke PC. Physical adsorption of charged plastic nanoparticles affects algal photosynthesis. J Phys Chem C 2010;114(39):16556−61. Available from: https://doi.org/10.1021/jp1054759.

[7] Yu F, Yang C, Zhu Z, Bai X, Ma J. Adsorption behavior of organic pollutants and metals on micro/nanoplastics in the aquatic environment. Sci Total Environ 2019;694:133643. Available from: https://doi.org/10.1016/j.scitotenv.2019.133643.
[8] Cunha C, Silva L, Paulo J, Faria M, Nogueira N, Cordeiro N. Microalgal-based biopolymer for nano- and microplastic removal: a possible biosolution for wastewater treatment. Environ Pollut 2020;263:114385. Available from: https://doi.org/10.1016/j.envpol.2020.114385.
[9] Ma B, Xue W, Ding Y, Hu C, Liu H, Qu J. Removal characteristics of microplastics by Fe-based coagulants during drinking water treatment. J Environ Sci 2018;78:267–75. Available from: https://doi.org/10.1016/j.jes.2018.10.006.
[10] Ma B, Xue W, Hu C, Liu H, Qu J, Li L. Characteristics of microplastic removal via coagulation and ultrafiltration during drinking water treatment. Chem Eng J 2019;359:159–67. Available from: https://doi.org/10.1016/j.cej.2018.11.155.
[11] Perren W, Wojtasik A, Cai Q. Removal of microbeads from wastewater using electrocoagulation. ACS Omega 2018;3(3):3357–64. Available from: https://doi.org/10.1021/acsomega.7b02037.
[12] Herbort AF, Sturm MT, Fiedler S, Abkai G, Schuhen K. Alkoxy-silyl induced agglomeration: a new approach for the sustainable removal of microplastic from aquatic systems. J Polym Environ 2018;26:4258–70. Available from: https://doi.org/10.1007/s10924-018-1287-3.
[13] Poerio T, Piacentini E, Mazzei R. Membrane processes for microplastic removal. Molecules 2019;24(22):4148. Available from: https://doi.org/10.3390/molecules24224148.
[14] CMT aids removal of microplastics from water streams. Membrane technology 2019;1:4–5. Available from: https://doi.org/10.1016/s0958-2118(19)30009-6 2019.
[15] Li L, Liu D, Song K, Zhou Y. Performance evaluation of MBR in treating microplastics polyvinylchloride contaminated polluted surface water. Mar Pollut Bull 2020;150:110724. Available from: https://doi.org/10.1016/j.marpolbul.2019.110724.
[16] Enfrin M, Dumée LF, Lee J. Nano/microplastics in water and wastewater treatment processes – origin, impact and potential solutions. Water Res 2019;161:621–38. Available from: https://doi.org/10.1016/j.watres.2019.06.049.
[17] Kumar RV, Kanna GR, Elumalai S. Biodegradation of polyethylene by green photosynthetic microalgae. J Bioremediat Biodegrad 2017;8:381. Available from: https://doi.org/10.4172/2155-6199.1000381.
[18] Klein S, Dimzon IK, Eubeler J, Knepper TP. Analysis, occurrence, and degradation of microplastics in the aqueous environment. Freshwater Microplas 2017;58:51–67. Available from: https://doi.org/10.1007/978-3-319-61615-5_3.
[19] Ariza-Tarazona MC, Villarreal-Chiu JF, Barbieri V, Siligardi C, Cedillo-González EI. New strategy for microplastic degradation: green photocatalysis using a protein-based porous N-TiO_2 semiconductor. Ceram Int 2018;45:9618–24. Available from: https://doi.org/10.1016/j.ceramint.2018.10.208.
[20] Kang J, Zhou L, Duan X, Sun H, Ao Z, Wang S. Degradation of cosmetic microplastics via functionalized carbon nanosprings. Matter 2019;1:745–58. Available from: https://doi.org/10.1016/j.matt.2019.06.004.

CHAPTER NINE

Removal of toxic algal blooms from marine water

9.1 Introduction

Harmful algal blooms (HABs) are natural phenomena that induce the proliferation of phytoplankton, thereby spreading throughout the coastal areas and estuaries. More areas are being affected by HAB, the development of toxic species and consequently heavy economic losses. These marine HABs are familiar plankton and are clearly visible because of their pigments at high biomass and their involvement in acute shellfish and fish poisoning of human and aquatic organisms. These blooms are found to be widespread due to anthropogenic activities, along with natural factors that are increasing the frequency of bloom and their toxins. There are nearly 150 HAB species known, which include dinoflagellates, diatoms, raphidophytes and prymnesiophytes, with fewer harmful species amongst the cyanobacteria and pelagophytes [1]. Marine algal toxins are also responsible for certain hazardous health effects in human and marine living organisms. The production of algal toxins occurs when unicellular algae in favourable conditions proliferate and aggregate to form dense concentration of cells or blooms. Phytoplankton species that produce toxins are categorized as HAB. There are two groups, dinoflagellates and diatoms, which produce toxins that are harmful to human health. Consumption of seafood contaminated with algal toxins can result in one of five seafood poisoning syndromes. They are paralytic shellfish poisoning, neurotoxic shellfish poisoning, amnesic shellfish poisoning, diarrhetic shellfish poisoning and ciguatera fish poisoning [2]. Though their impacts are not concentrated, their removal from the environment is needed. This chapter covers some of the health impacts and toxin removal methodologies in detail.

Modern Treatment Strategies for Marine Pollution.
DOI: https://doi.org/10.1016/B978-0-12-822279-9.00001-4

9.2 Harmful effects of harmful algal bloom

Harmful algal blooms can occur in water at any time due to excess accumulation which may impair water use. There are lots of harmful effects that are witnessed with regard to ecologic, economic and public health concerns [3].

9.2.1 Ecologic concerns

- Decrease in oxygen demand
- Disruption in food web
- Movement of sunlight is disrupted [3]

9.2.2 Economic concerns

- Increases the drinking water cost
- Loss of recreational revenue
- Loss of taste and odour or alteration [3]

9.2.3 Public health concerns

- Induction of allergic reaction
- Toxicity produced by cyanobacterium [3]

Some of the algal species that cause pollution are *Euglena*, diatoms, golden algae, cyanobacterium and green algae. These algal blooms commonly occur in the spring and winter which favours algal growth due to increased moisture in air. The major factors influencing algal bloom occurrence are [3]

- nutrients;
- water clarity;
- hydrology, climate and weather conditions;
- biological community interactions;
- phosphorous and nitrogen [3].

Eutrophication is a process of growth of unwanted algae in water which leads to visible algal blooms that cause an increase in turbidity, thereby affecting odour and taste problems. During algal blooms toxins are produced that render water unsafe and cause fish mortality, which affects humans in terms of food consumption. Diatoms and dinoflagellates are involved in the production of toxins responsible for numerous poisonings in humans which affect mainly the nervous and the intestinal system. Sectors like fisheries,

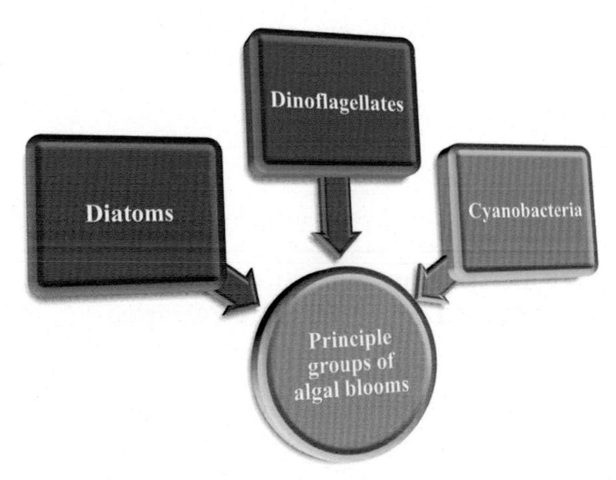

Figure 9.1 Principle algal groups that form blooms.

tourism and recreation are severely affected [4]. Fig. 9.1 gives an idea on groups of algal blooms available in environment.

9.3 Algal blooms and its toxins

Favourable environmental and climatic conditions can prevail to induce HABs, which have the capacity to produce hazardous toxins that find their way through the food chain. Different physical, chemical and biological factors are responsible for HAB appearance and the resulting toxins production. These factors include rise in temperature, an increase in rainfall events and excessive nutrient discharge into water bodies. Toxins are released when the algal bloom dies off. However toxins are released into the water by live algal cells. The cell membrane rupture causes the spread of the toxins in water bodies and their absorption by organic material in the water column. Marine toxins have been classified into five groups according to the effects they cause to organisms: paralytic shellfish poisoning (PSP), diarrhetic shellfish poisoning (DSP), amnesic shellfish poisoning (ASP), neurotoxic shellfish poisoning (NSP) and azaspiracid shellfish poisoning (AZP). ASP is caused by the ingestion of toxins released by diatoms, whereas others are released by dinoflagellates which contaminate shellfish. Cyanobacteria are responsible for HABs in freshwater and to a lesser extent in marine environments. They are able to produce toxins which are classified as: (1) hepatotoxins; (2) cytotoxins that produce both hepatotoxic and nephrotoxic effects; (3) neurotoxins and (4) dermatoxins that cause irritant responses on contact [4].

9.4 Mechanism involved in occurrence of harmful algal bloom and its removal strategies

Considering the toxicity effects of algal blooms and their toxins on aquatic and human health many physical—chemical treatments are suggested at the small-scale/laboratory level. Generally these harmful organisms are removed from freshwater using technologies, such as adsorbents like activated carbon, degradation like photolysis, the use of catalysts in water like permanganate and hydrogen peroxide and disinfection using ozonation and chlorination. Additionally reverse osmosis has been used to remove these toxins from water [5]. This chapter covers major three removal technologies: the application of phosphatic clay, the influence of metal ions and use of polymer flocculants.

The occurrence, abundance and geographical distribution of algae and their toxins or cyanobacterial blooms have increased substantially due to the increased anthropogenic input of organic matter and nutrients and also because of global warming. They are responsible for physiological, ecological and environmental adverse effects [6]. They are:

- alteration of water quality with high eutrophication;
- depletion of dissolved oxygen below pycnocline;
- loss of seagrass and benthos;
- loss of phytoplankton competitor motility;
- inhibition of enzyme and photosynthesis;
- cell and membrane damage;
- mortality of fish, coral reef, livestock and wildlife;
- shellfish or finfish poisoning caused by neurotoxic compounds produced by red tide dinoflagellates;
- illness or even death of higher organisms including humans and animals that are associated with the consumption of contaminated fish, seafood and water;
- adverse health effects (e.g., eczema or acute respiratory illness) from direct contact with, ingestion, or inhalation of cyanobacteria or various toxins, during recreational or occupational activities [6]

9.4.1 Mechanism behind formation of harmful algal bloom

The main underlying cause of the increasing occurrence of harmful algal blooms is global warming on water with a high content of dissolved organic matter (DOM) and particulate organic matter (POM, e.g., phytoplankton or algae) via the photosynthesis process. The main reason behind the formation of HAB is the presence of organic matter that induces the fuel production, nutrients and various photochemical and microbial products [6].

$$\text{DOM} + \text{POM} + \text{hn} \rightarrow H_2O_2 + CO_2 + \text{DIC} + \text{LMW DOM} + NO_3 + NH_4^+ + PO_4^{3-} + \text{autochthonous DOM} + \text{other species}$$

This is a complex photoinduced reaction yielding nutrients for algae. DIC is defined as the sum of an equilibrium mixture of dissolved CO_2, H_2CO_3, HCO_3^-, CO_3^{2-} and low molecular weight (LMW) DOM formed through the photonic breakdown of larger organic matter [6].

The microbial reaction involved in yielding nutrients and final products in water is given below.

$$\text{DOM} + \text{POM} + \text{microbes} \rightarrow H_2O_2 + CO_2 + \text{DIC} + \text{LMW DOM} + NO_3^- + PO_4^{3-} + NH_4^+ + \text{autochthonous DOM} + \text{other species } [20]$$

The compounds formed from DOM and POM through microbial and photochemical processes will substantially increase because of the increased temperature from global warming. This phenomenon is found to be the main reason for the increasing incidence of harmful algal blooms in water with huge content of DOM and POM, leading to eutrophication in the presence of light. Another possible reason is the regeneration of autochthonous DOM and nutrients from POM in DOM rich water that serves as a primary food for HAB. In contrast global warming affects water with a low content of DOM in the opposite direction by inhibiting the production and regeneration of various compounds found in water. As a consequence this limits the growth of algal blooms by limiting the photosynthetic process [6]. These are the reasons for growth of harmful algal blooms in water and other bodies.

9.4.2 Solutions for algal bloom control

It is necessary to add remedial measures to control the growth of algal blooms in freshwater and coastal water systems. Basically prevention measures are to be taken to avoid eutrophication. Controlling the release of organic matter inputs like DOM and POM can reduce the regeneration of photoproducts, microbial products and nutrients. This may probably reduce photosynthesis in the water system. Additionally, the removal of algae or phytoplankton during algal blooms using fine, small mesh nets and its sediments would reduce the photoinduced and microbial release of DOM and nutrients from primary production. Finally, treatment protocols can be adopted for removing the algae blooms from water. Some of the treatment procedures/methodologies are discussed at the lab scale because they are proposed as a solution in research for removing or recovering the natural water systems. Knowledge and sophistication should be considered for large-scale applications [6].

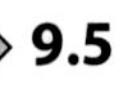

9.5 Removal of red tide organism using flocculants

9.5.1 Cationic polymeric flocculant

Harmful algal blooms are called red tide organisms due to the mark made by the discoloration of seawater surface due to the rapid growth and accumulation of certain microalgae. Numerous technologies like inducing algal flocculation and rapid settling from the water column have been proposed to remove algae from water. This flocculation is carried out by adding chemicals like alum. Economically clay has been proposed as a flocculant but it can induce rapid dilution. With the characteristics of nontoxicity and biodegradability, polymers like amylopectin, starch, cellulose and lignin have been used. This polymer source is grafted into cationic polymers to perform efficient flocculation processes. Though they are synthetic flocculants, they show better results in completely eliminating HAB from marine bodies [5]. Pang et al. [5] fabricated natural cationic polymeric flocculants grafted by quaternary ammonium monomer *N*-(3-chloro-2-hydroxypropyl) trimethyl ammonium chloride (CHPTAC) onto the backbone of corncob powder. Removal of HAB in seawater was 70% in 6 h which clearly showed that it can be used as a promising substrate for reducing adverse effects from harmful algal blooms in seawater.

9.5.2 Clay

Clay with naturally occurring silica has been used as a flocculant to remove red tides from seawater. Use of clays to eradicate an existing HAB is a promising and attractive direct control option for locations with persistent HAB problems. Clays are appropriate for this purpose because they are generally inexpensive, readily available in large quantities and easy to use in field operations. In addition, clays are thought to be substances with little or no direct toxic effects on aquatic organisms. Marine organisms can adapt to varying amounts of clays but the impacts on them are still unknown. Some of them like montmorillonite, bentonite and Florida phosphatic clay displayed removal efficiencies greater than 90%. The addition of clay on algal cultures (Laboratory scale) yielded cell death, but it was later found that physical contact between clay and cell yielded death not because of toxins from clay particles. They work under the principle of the coagulation and flocculation mechanism. Clay forms flocs with algae that are separated out by settling [7].

9.5.3 Composite clay

Natural clay has a negatively charged surface that tends to have a poor capacity to flocculate. This drawback is addressed by using composite clays like quaternized chitosan and montmorillonite. Quaternized chitosan intercalated montmorillonite effectively removed HAB by the deposition netting and electrostatic neutralization mechanism, thereby clearing cyanobacterial bloom in a natural water body within 48 h. The quaternized chitosan acts as an algaecide in the interlayer of montmorillonite that inhibits the growth and resuspension of algae aggregates [8].

9.5.4 Composite sand

Chitosan-modified sand with flocculants like *Moringa oleifera* should increase removal efficiency as it altered the isoelectric point of sand. High HAB removal rate is achievable when the sand is modified by the bicomponent mechanism of surface charge and netting-bridging modification using biodegradable modifiers such as MO and Chitosan. This removal is achieved by the electrostatic interaction between the algal cell and flocculant [9]. Fig. 9.2 illustrate coagulation and flocculations as major mechanism in HAB removal using sand and its composites.

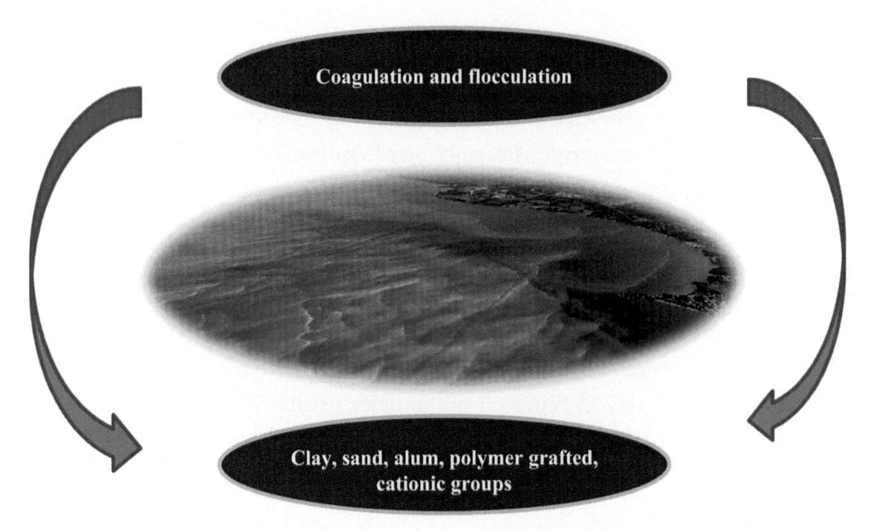

Figure 9.2 Removal of HAB using flocculation. *HAB*, Harmful algal bloom.

9.6 Influence of low zinc, copper and iron ions in limiting bloom growth

9.6.1 Effects of low concentration of trace metal in aquatic environment

The total dissolved concentration of dissolved zinc in the surface water of the open Pacific and Atlantic oceans is around 0.1 nmol/L. Its availability in water is reduced by the presence of organic ligands. It is estimated that the availability of Zn^{2+} reduces the phytoplankton growth. Zinc acts as a cofactor in carbonic anhydrase, an enzyme that enhances the rate of dehydration of bicarbonate ions to free carbon monoxide. It has an important role in diatoms like silicification [10]. Generally total zinc concentration in unpolluted marine water is in the 10^{-10} M range. Up to now it was accepted that only iron is considered as growth-limiting micronutrient in the marine environment. It is predicted from the research work that the toxicity of a trace metal depends on the chemical speciation that is related to the free ion activity. Unlike other metals like copper and iron, zinc is considered to be a cofactor for many enzymes like dehydrogenase, carbonic anhydrase and carboxypeptidase. Under limited concentrations they tend to limit the growth of algae in water [11].

The basic concentration of copper and zinc ranged from 10^{-14} to 10^{-16} M and from 10^{-8} to 10^{-9} M, respectively. Basically there are lots of differences between freshwater and marine water algae. Freshwater algae show a higher tolerance to these metals than marine algae. Freshwater algae showed a higher tolerance to copper than to zinc; this is because they tend to immobilize it intracellularly. This shows they have a high affinity for copper ions, but this is not the case in marine algae. They need a minimum concentration of these metals for its proliferation and multiplication [12].

9.6.2 Trace metals serve as limiting nutrients

Trace metals like Fe, Zn, Co and Cu play vital roles in the growth and metabolism of all aquatic algae. They help in the photosynthesis and assimilation of other nutrients. Iron helps metabolic functions like photosynthetic electron transport, respiration, nitrate assimilation, N_2-fixation and detoxification of reactive oxygen species. Because of its heavy involvement in photosynthetic electron transport, cellular iron requirements increase with decreasing light intensity. Zinc has a variety of metabolic functions that occur in carbonic anhydrase (CA), an enzyme critical to CO_2 transport and fixation. Higher amounts of this enzyme are needed under CO_2-limiting conditions, hence the importance of zinc increases. It also occurs in zinc finger protein that is involved in DNA transcription and to acquire phosphorous from organic phosphate esters. Copper, which is important for respiratory electron transport, occurs in cytochrome oxidase. It is an important component of the high-affinity iron transport system of many eukaryotic algae. Other metals like nickel and molybdenum are important in nitrogen assimilation [13].

9.6.3 Toxicity effects of trace metals

More reactive metals like Co, Zn and Cu are toxic at higher concentrations. Growth of algae is limited under limited concentration, whereas growth is inhibited at higher concentration. Toxicity effects are taken up through the transport system of nutrient metals. Algae are highly sensitive to the varied concentration of copper ions in water. Algal grazers like copepods and ciliates are more sensitive to trace metals like Cu and Zn than phytoplankton. It also has some effects on the diatom genus *Pseudonitzschia.* These algae produce the neurotoxin domoic acid, which has caused poisoning in marine mammals, seabirds and humans. Generally they appear in iron-limited oceanic upwelling systems and growth is stimulated by iron addition in the mesoscale, as found in laboratory experiment. It is found from lots of experiments that by promoting iron uptake, domoic acid production could convey a competitive advantage to toxigenic *Pseudonitzschia* species in iron-limited oceanic and coastal upwelling systems. It also had toxicity effects in other HAB by interrupting the nitrogen fixation process [13].

9.7 Removal using coagulation—magnetic separation method

There are various methods for removing algal blooms and their toxins from aqueous bodies. This is because algal blooms seriously damage the balance and stability of aquatic ecosystem. Algal blooms involving toxin-producing species that possess serious threats to animals and human. Among the various removal technologies cleaning using clay coagulant is gaining global popularity. Though clay is cheap and economical it has a few limitations like it causes excessive sediment siltation and considerable dredging work [14]. Hence focus has switched over to fly ash waste coming from thermal power plants. Fly ash was compounded with magnetite to incorporate a magnetic property in order for its use as a coagulant.

Research by Lui et al. [14] states that the coagulation—magnetic separation technique is highly effective in clearing up harmful algal blooms from water and mitigate eutrophication. A magnetic coagulant was synthesized by compounding acid-modified fly ash with magnetite. After mixing, coagulation and magnetic separation, the flocs are obtained from the magnet surface. It is examined that more than 99% of algal cells were removed within 5 min after the addition of magnetic coagulant at optimal loadings. It also showed good removal efficiencies in clearing COD, total nitrogen and phosphorus. The main mechanism behind removing algal blooms from water is mesoporous adsorption, netting and bridging and high magnetic responsiveness to a magnetic field. This method shows good performance by being low cost, since it turns waste into some valuable compounds that can be used in many pilot plants for the future prevention of algal contaminations [14].

9.8 Microorganism-based control of algal blooms

As problems induced by harmful algal blooms to the environment have gained attention, many removal methods have been proposed, e.g., physical, chemical and biological. Though physical and chemical methods show good efficiency in cleaning algal blooms from water they possess certain disadvantages like high cost, secondary pollution and nontarget toxicity to aquatic organisms and human. Thus the development of biological methods has been initiated. Biological methods for the removal of harmful algal blooms include the use of aquatic plants, aquatic animals and algicidal microorganisms to limit the growth of algae. Among these methods microorganisms pave the way for effective cleaning of algal blooms because of their potential equal to algicides and their environmental friendliness. Microorganisms-based methods use algicidal bacteria, fungi, viruses and protozoa and are considered promising eco-friendly approaches, compared to the physical and chemical methods [15].

Microorganism-based methods are classified into two types: single species methods in which pure strains are responsible for the control of harmful algal blooms and microbial aggregates methods in which multi-species communities are responsible for algal growth control. It is reported that microorganism are used in the form of flocculants to effectively control harmful algal blooms in water within a short span of time. Apart from chemical flocculants, bioflocculants are the most promising candidate in the field of coagulation [15].

9.8.1 Based on bacterial bioflocculation

Bacteria play important roles in the blooming and dying out of harmful algae. Bacteria promote the aggregation of algal cells by (1) inducing the release of extracellular polymeric substances (EPS) by algae, thus accelerating the aggregation of algal cells and (2) producing bioflocculants that directly cause the accumulation and settling of algal cells. Bacterial bioflocculants are identified as polysaccharides, protein and lipids that are secreted outside the bacterial cell. Bioadsorption, ion bridging and charge neutralization are the main mechanisms for bacteria-associated bioflocculation. This can be facilitated by proteins and flagella on the cell wall [15].

9.8.2 Based on fungal bioflocculation

Removal of algal cells from water is assisted by pelletization formed by fungi. The removal of algal cells by the interaction of algal cells with fungal cells during the pelletization process of fungi can result in the formation and precipitation of fungus–alga pellets, removing algae from water bodies. The formation of fungus–alga pellets starts with swelling and germinating spores, followed by growth and branching of hypha. Factors that influence pelletization can be physicochemical properties of the spores and hyphae and the cultivation parameters, such as pH values, salinity and rheological behaviour of the growth medium. Charge-assisted filamentous fungi can attract algal cells by changing their surface charge. The pellets survive through fungus–alga interaction; the algae fix CO_2 and produce organic compounds for fungal growth, while fungi entrap the algae by hyphae production. After some period of time autolysis of fungi cell occurs which results in dismantling the pellets. Hence it shows that the life span of pellets depends on the life span of the fungi only. They are employed as bioflocculants in removing harmful algal cells from water [15].

9.8.3 Removal/killing mechanism of microbes on harmful algal blooms

The main mechanism involved in inhibiting the harmful growth of algal is surviving competition. Some bacterial strains inhibit the growth of algae in lakes through the secretion of allelochemicals. It is found from the research that bacterial crude extract β-carbolines showed antagonistic activity against cyanobacteria through algal cell division. Similarly fungi also showed some inhibitory effects against algal growth. It is reported that a white-rot fungus *Lopharia spadicea* showed significant inhibitory effects on *Microcystis aeruginosa*, *Glenodinium* sp. and *Cryptomonas ovate* [15].

Algicidal bacteria are an agent involved in killing algae found in normal lake or aquatic ecosystems. Algicidal bacteria kill through direct or indirect contact with algae. *Myxobacteria* were the first reported algicidal bacterium, which could kill unicellular and filamentous *Cladophora* on close contact. Mostly these microbes kill by secreting algicidal compounds. The algicidal active substances excreted by bacteria are compounds that include peptides, proteins, alkaloids, amino acids, antibiotics, pigments and fatty acid. The same mechanism applies for fungi, actinomycetes and other microbes [15].

9.8.4 Microbial aggregates for algal growth control

Microbial aggregates are embedded in a mucilage matrix of an extracellular polymeric substance that has high stability and cell density. They are divided into two types. They are heterotrophic biofilms and phototrophic biofilms. Phototrophic biofilms are called periphyton of microbial aggregates that usually consist of algae, bacteria and meso/microorganism. They spread between overlying water and sediments, specifically on the surface of sediments, rocks, plants and suspended particles in aquatic ecosystems. They inhibit the growth of algae by generating allelochemicals. Heterotrophic biofilms are also called activated sludge biofilms. Generally microbial aggregates control algal growth through adsorption–desorption, activation–transformation, absorption–metabolism, biological nitrogen fixation and changing the environmental conditions like pH [15].

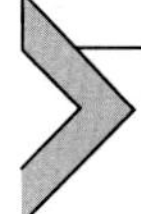

9.9 Control of algal growth through ultrasound technology

Ultrasound is sound at frequencies higher than those that are detected by the human ear. The frequency range of 20–200 kHz is applied in medicine, science, industrial processes and water treatment methodologies. Research has found that ultrasound can be effectively applied to clear algal blooms and their toxins from water bodies. They can collapse the gas vesicles, disrupt the cell wall, disturb photosynthetic activity and finally cause cell lysis of algal blooms [16]. Ultrasound radiation reduces the algal growth by structural or functional destruction. This is done through the generation of cavitation bubbles which collapse and cause localized temperature and pressure to reach 5000°C and 500 atmosphere of pressure. This altered environment can destroy the buoyancy of algae by collapsing gas vacuoles, inhibiting the photosynthetic process and destroying the cell membrane of algal cells. Algal cells in water experience intense shockwaves and shear force caused by the collapse of acoustic cavitation bubbles. The destruction of water vapour with the collapsed bubble into free radicals can also destroy the cell wall through chemical reactions resulting in the loss of photosynthetic ability [17]. There are a few factors that affect the efficiency of operation, yielding good destruction of cells. They are frequency of ultrasonic waves, intensity of power and duration of exposure to sonication. It is reported that frequencies applied for the destruction of algal cells are in the range 20–1144 kHz. The use of lower frequencies is adopted because it requires a lower electric current, therefore it is lower cost. But higher efficiency is observed in higher frequencies for controlling the growth of algae. Higher-frequency sonication also tends to produce more free radicals which can damage algal cells. But a higher frequency also requires more electric power to generate cavitation. The required exposure time of sonication for controlling algae varies depending on ultrasonic frequencies and intensities, algal biomass concentrations and distribution and environmental conditions. Generally exposure time is proportional to the degradation of algal biomass [17].

It is observed from studies that 5 min of sonication at 20 kHz with a higher intensity of 0.32 W/cm^3 resulted in more than 60% of *M. aeruginosa* removal. It is found that there is a significant increase in microcystin content with a decreased exposure of time. Further findings state that ultrasonic irradiation shorter than 5 min was known to be effective for the inhibition of algal growth, while not releasing cyanotoxins from the algal cells. This shows that the duration of exposure to ultrasound is an important factor to consider for the minimal release of algal toxins during the degradation process [17]. Fig. 9.3 shows some of the factors affecting killing of algal blooms in sonication.

When this technology is applied in the field certain factors are to be considered. Field environmental conditions such as rainfall and water quality parameters affect algal biomass concentration, photosynthetic

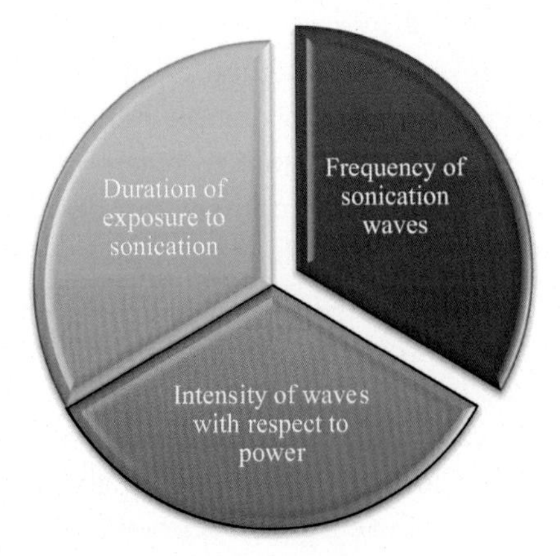

Figure 9.3 Factors influencing destruction of algal blooms using sonication.

ability and efficiency of ultrasound during the test time. The distribution of algae in the water depends on water flow, wind and turbulence in water. Such distribution of algal biomass can be witnessed using a combination of remote-sensing applications. Remote sensing enables one to sense rapid occurrence of algae in water bodies through which water treatment using ultrasound waves can be established on-site under real-time conditions. There are many advances in the sonication process like hydraulic jet cavitation to destroy gas vesicles for algal removal and hydrodynamic cavitations to break the growing conditions of algae have been applied and studied in recent times [17].

9.10 Triboelectric nanogenerator for algae removal using water wave energy

Ambient water motions like ocean tides, waves and seawater currents is a gigantic source of renewable mechanical energy. Triboelectric nanogenerator (TENG) is a creative method invented for converting ambient mechanical energy into electricity through the periodic contact and separation of two polymer plates. TENG is also driven by the kinetic energy of water motions that gives the possibility of fabricating self-powered environmental protection system driven by water motions. It is made of organic material that uses the electrification effect [18]. When TENG is coupled with a seawater electrolysis device it introduces a new type of electrochemical water treatment system. In this system reduced graphene oxide on the titanium oxide is used as an electrode for the lysis of algal cells and for disinfecting water. This system can successfully work under the power provided by the fabricated TENGs without an external power source, which leads to the invention of a totally self-powered, self-contained active electrochemical water treatment system for sterilization and algae removal [18].

This new electrochemical water treatment system consists of TENG as a power supply and a reactor with water and electrode plates. The experimental setup has both cathode and anode placed at a distance of 2 cm. The MMO−Ti−RGO electrode plates were made from commercial MMO−Ti plates by electrodepositing reduced graphene oxide (rGO) on the surface. Graphene is a two-dimensional (2D) hexagonal lattice structure of sp^2 hybridized carbon atoms, which can suppress the proliferation of microbes on its surface even at environmentally suitable conditions. All microbes including algae in the electrode plates are electroporated by a strong electric field. Along with this, chlorine is helpful in removing all microbes like bacteria and algae from water. The system takes advantage of 2D nanomaterials and seawater electrolysis to achieve high efficiency disinfection for both bacteria and algae. This system displays immediate practicability for treating the polluted ocean [18].

9.11 Electrochemical treatment for removal of algae from water

Though different process like coagulation, floatation and sand filtration are proposed for algae removal from natural waterways and high removal yields with aluminium salts for different marine microalgae are in use or discussed in many literature, it can be witnessed often that these treatments are coupled with oxidative treatments that incorporate chemical oxidizing agents. These chemical agents include ozone, chlorine dioxide, chlorine or permanganate. They are produced electrochemically and sent for treatment purposes. A combination of electrochemical processes with disinfection units are used for removing organic matter, implying its efficiency in cleaning algae from water. The main mechanism involved in the inactivation of microorganisms in electrolysis is due to the action of free radicals, long-life oxidants that are electrogenerated and the effect of the electric field on the cell membrane [19]. In this process electrodic generation of free radicals is needed for killing algal cells. For this purpose electrodes are used for generation of oxidants in water. In particular electrodes coated with boron-doped diamond (BDD) have characteristics, such as chemical inertness and high overvoltages for oxygen evolution, which can be suitable for their use in water treatment. This process mainly depends on current density, concentration of target pollutants in water, distance between electrodes and chloride or other oxidant concentration in water. The optimum current density is in the range of 25–100 Am^{-2} [19].

A commercial filter-press electrochemical reactor (electro MP-cell), equipped with a BDD anode and a stainless steel cathode, was used to perform the electrolysis. The cell was inserted in a hydraulic circuit and used in either a closed loop, as a recirculating batch reactor, or in the continuous mode. The study demonstrated the effect of current density and hydrodynamics on the formation of active chlorine and other chloride oxidation products. At 100 mg/dm^3 of chlorides complete removal of algae was observed in water. The experiment was tested with green algae (*Chlorella vulgaris*). This shows that algae removal can be achieved using long-life oxidants that are electrogenerated in water using electrolysis [19]. The main mechanism involved in the electrolysis of water containing chloride at the BDD anode is

$$H_2O \rightarrow OH^- + H^+ + e^-$$

$$Cl^- \rightarrow \frac{1}{2} Cl_2 + e^-$$

$$Cl_2 + H_2O \rightarrow HClO + H^+ + Cl^- [22]$$

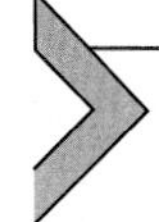

9.12 Miscellaneous removal of algal blooms and toxins from water

9.12.1 Using seawater reverse osmosis technology

Seawater distillation serves numerous populations for drinking water purposes. It is performed using a reverse osmosis process. There are various methods of seawater distillation, like thermal distillation and reverse osmosis. Reverse osmosis is more energy efficient, more compact and more flexible compared to other desalination processes. Generally water production cost is lower in reverse osmosis. One of the major drawbacks is membrane fouling that affects the efficiency of the removal of salts from water. Although 99% of algal removal is achieved in microfiltration and ultrafiltration, the membrane can become clogged in membrane [20]. The excessive biomass and organics associated with HAB can lead to the closure of desalination plants, particularly seawater reverse osmosis (SWRO) plants due to overloading of the pretreatment facilities or potential irreversible RO membrane fouling. This technology can be used for removing algal toxins by predicting the toxin levels at the intake and outside of the process, especially in reverse osmosis, because in thermal distillation due to the high molecular weight and high boiling point organics will remain in brine. In reverse osmosis they are prone to be removed through size exclusion mechanism [21].

9.12.2 Using vermiculite

Vermiculite is a magnesium silicate mineral with various amounts of iron and aluminium, which is chemically similar to mica and montmorillonite. It is positively charged which is enhanced by the tetrahedral substitution of aluminium and iron for silica which is improved by the presence of interlayer cations. Vermiculite and vermiculite modified with hydrochloric acid were investigated to evaluate their flocculation efficiencies containing harmful algae blooms. It was found that the vermiculite modified with hydrochloric acid could coagulate algae cells through charge neutralization, chemical bridging and netting effect. Nearly 88% of algae cells are removed from the solution using modified vermiculite [22].

9.13 Conclusion

The treatment methods reviewed here are coagulation and flocculation using some natural clays, polymers and composites and reverse osmosis technology. These technologies are available for removing algal blooms from water. Algal blooms are toxic to the marine environment and to human health. The level of toxicity is influenced by various nutrients, water clarity, hydrology, climate and weather conditions, biological community interactions and phosphorous and nitrogen content. Though there are lots of technologies available on the market for detecting and monitoring these algal blooms, there are fewer methods for their removal. Removal through conventional treatment units like sedimentation and filtration may increase their efficiency but decrease the survival rate of those cleaning units. It is mandatory to prevent the growth of blooms in water by reducing the growth-promoting factors like the influence of metal ions and the presence of nutrients. Clearing algal blooms using novel technologies has to be encouraged using research ideas. Instead of treating algal bloom-contaminated water, the prevention of its growth through controlled discharge of anthropogenic activities in water and other related issues has to be achieved. Prevention is better than cure is the main theme in the case of algal bloom toxicity effects.

References

[1] Landsberg J, Van Dolah F, Doucette G. Marine and estuarine harmful algal blooms: impacts on human and animal health. Oceans and health: pathogens in the marine environment. 2005. p. 165–215. Available from: https://doi.org/10.1007/0-387-23709-7_8.

[2] Van Dolah FM. Marine algal toxins: origins, health effects, and their increased occurrence. Environ Health Perspect 2000;108(Suppl. 1):133–41. Available from: https://doi.org/10.1289/ehp.00108s1133.

[3] <www.epa.gov/production/files/documents/harmfulalgalbloom.pdf>

[4] <https://publications.jrc.ec.europa.eu/repository/bitstream/JRC101253/lbna27905enn.pdf>

[5] Pang Y, Ding Y, Sun B. Removal of red tide organism by a novel cationic polymeric flocculant. Procedia Environ Sci 2013;18:602–9. Available from: https://doi.org/10.1016/j.proenv.2013.04.083.

[6] Mostofa KMG, Liu C-Q, Vione D, Gao K, Ogawa H. Sources, factors, mechanisms and possible solutions to pollutants in marine ecosystems. Environ Pollut 2013;182:461–78. Available from: https://doi.org/10.1016/j.envpol.2013.08.005.

[7] Sengco MR, Li A, Tugend K, Kulis D, Anderson DM. Removal of red- and brown-tide cells using clay flocculation. I. Laboratory culture experiments with Gymnodinium breve and *Aureococcus anophagefferens*. Mar Ecol Prog Ser 2001;210:41–53.

[8] Gu N, Gao J, Wang K, Zhao Y, Li H, Ma Y. Quaternized chitosan-intercalated montmorillonite composite for cyanobacterial bloom inhibition. Desalination Water Treat 2015;57(42):19665–76. Available from: https://doi.org/10.1080/19443994.2015.1106350.

[9] Li L, Pan G. A universal method for flocculating harmful algal blooms in marine and fresh waters using modified sand. Environ Sci Technol 2013;47(9):4555–62. Available from: https://doi.org/10.1021/es305234d.

[10] Crawford DW, Lipsen MS, Purdie DA, Lohan MC, Statham PJ, Whitney FA, Putland JN, Johnson WK, Sutherland N, Peterson TD, Harrison PJ, Wong CS. Influence of zinc and iron enrichments on phytoplankton growth in the northeastern subarctic Pacific. Limnol Oceanogr 2003;48(4):1583–600.

[11] Anderson MA, Morel FMM, Guillard RRL. Growth limitation of a coastal diatom by low zinc ion activity. Nature 1978;276(5683):70–1. Available from: https://doi.org/10.1038/276070a0.

[12] Knauer K, Behra R, Sigg L. Effects of free Cu^{2+} and Zn^{2+} ions on growth and metal accumulation in freshwater algae. Environmental Toxicology and Chemistry 1997;16(2):220–9.

[13] Sunda WG. Trace metals and harmful algal blooms. In: Ecology of Harmful Algae; n.d. p. 203–214. https://doi.org/10.1007/978-3-540-32210-8_16.

[14] Liu D, Wang P, Wei G, Dong W, Hui F. Removal of algal blooms from freshwater by the coagulation–magnetic separation method. Environ Sci Pollut Res 2012;20(1):60–5. Available from: https://doi.org/10.1007/s11356-012-1052-4.

[15] Sun R, Sun P, Zhang J, Esquivel-Elizondo S, Wu Y. Microorganisms-based methods for harmful algal blooms control: a review. Bioresour Technol 2018;248:12–20. Available from: https://doi.org/10.1016/j.biortech.2017.07.175.

[16] Lürling M, Waajen G, de Senerpont Domis LN. Evaluation of several end-of-pipe measures proposed to control cyanobacteria. Aquat Ecol 2015;50(3):499–519. Available from: https://doi.org/10.1007/s10452-015-9563-y.

[17] Park J, Church J, Son Y, Kim K-T, Lee WH. Recent advances in ultrasonic treatment: challenges and field applications for controlling harmful algal blooms (HABs). Ultrason Sonochem 2017;38:326–34. Available from: https://doi.org/10.1016/j.ultsonch.2017.03.003.

[18] Jiang Q, Jie Y, Han Y, Gao C, Zhu H, Willander M, et al. Self-powered electrochemical water treatment system for sterilization and algae removal using water wave energy. Nano Energy 2015;18:81–8. Available from: https://doi.org/10.1016/j.nanoen.2015.09.017.
[19] Mascia M, Vacca A, Palmas S. Electrochemical treatment as a pre-oxidative step for algae removal using *Chlorella vulgaris* as a model organism and BDD anodes. Chem Eng J 2013;219:512–19. Available from: https://doi.org/10.1016/j.cej.2012.12.097.
[20] Villacorte LO, Tabatabai SAA, Dhakal N, Amy G, Schippers JC, Kennedy MD. Algal blooms: an emerging threat to seawater reverse osmosis desalination. Desalination Water Treat 2014;55(10):2601–11. Available from: https://doi.org/10.1080/19443994.2014.940649.
[21] Boerlage S, Nada N. Algal toxin removal in seawater desalination processes. Desalination Water Treat 2014;55(10):2575–93. Available from: https://doi.org/10.1080/19443994.2014.947785.
[22] Miao C, Tang Y, Zhang H, Wu Z, Wang X. Harmful algae blooms removal from fresh water with modified vermiculite. Environ Technol 2013;35(3):340–6. Available from: https://doi.org/10.1080/09593330.2013.828091.

CHAPTER TEN

Miscellaneous technologies for removing micropollutants from marine water

10.1 Introduction

Water plays a crucial role in maintaining sustainability among living beings through balanced ecological cycle on Earth. The challenge lies in accessing clean water for industrial and domestic use. Although Earth is 78% covered with seawater only 0.003% is accessed by human directly. The need for water for daily and day to day use has been increasing due to increased demands in industrial and domestic sectors. The demand for clean water is increasing. Hence there are proposals to use seawater for man's needs. Lots of desalination plants have been initiated to meet the growing demand for water. Desalination is one of the emerging technologies with lots of innovations in predicting and fulfilling the demand acquired by growing population. Since the marine environment by itself is contaminated by lots of floating, suspended and dissolved debris care should be first taken in clearing all these pollutants from water using novel methods like microbial degradation and catchers for removing plastics from water. Apart from clearing these pollutants from water by using separate technologies, they can be cleared at the pretreatment stage used in the water treatment or desalination processes. Generally the desalination process has three stages: pretreatment, reverse osmosis and disinfection [1,2]. This is an idea that has been recently applied to the removal of algae and its toxins from marine water. This chapter covers some of the new innovations that are highly helpful in cleaning marine water that is polluted with numerous contaminants.

Modern Treatment Strategies for Marine Pollution.
DOI: https://doi.org/10.1016/B978-0-12-822279-9.00002-6

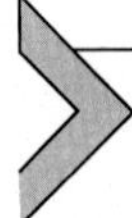

10.2 Nanomembranes—removing contaminants from seawater

10.2.1 Use of seawater in desalination process—as feed

Nanotechnology plays a major role in the purification of water and wastewater. Marine water has been used as a feed material in the desalination process to obtain clean and clear water for domestic usage. Generally reverse osmosis and thermal distillation are the two technologies that are used for the removal of salts from seawater. Among those techniques reverse osmosis is commonly used because of its efficiency in removing salts and cost effectiveness. Reverse osmosis uses a membrane in removing contaminants [3]. These membranes can be remodified using nanomaterials in order to enhance the removal of dispersed contaminants like algal toxins, micropollutants, heavy metals, xenobiotic compounds and many more. These pollutants are not of concern in conventional treatment systems. Nanotechnology aids the smart/advanced materials/membrane design through fabrication of 1D, 2D and 3D nanomaterials in order to bring revolutionary changes to desalination/water purification [4]. Versatile characteristics of nanotechnology in the fabrication of organic or inorganic-based nanomaterial membranes have helped to yield membranes with controlled shapes and pores distribution, improved porosity, surface area, volume ratio and optimized dimensions/shapes [4].

10.2.2 Methods adopted for fabricating nanomembranes

There are different techniques used for synthesizing nanomembranes, such as phase inversion, interfacial polymerization, track-etching and electrospinning [4].

10.2.2.1 Phase inversion

Chemical stratification is used to remove the solvent from the liquid–polymer solution and converts the homogeneous solution into a porous solid membrane. Membrane morphology and porosity are controlled by the nature of the solvent used in this process. Phase inversion is used to yield superoleophobic poly(acrylic acid)-graft PVDF membrane that induces efficient desalination and major emulsion/oil–water separations [4].

10.2.2.2 Interfacial polymerization

This is a process of step-growth polycondensation that occurs in immiscible solvents. This technique is used to fabricate membranes of thickness of 10 nm to microns that can be used for reverse osmosis (RO) and nanofiltration (NF) membranes [4].

10.2.2.3 Track-etching

This process involves energetic heavy ions irradiation onto a substrate that results in a linear damaged track across the irradiated polymer surface, thereby producing nanomembranes. The pore size distribution is nm to μm [4].

10.2.2.4 Electrospinning

This uses a high-voltage treatment to polymers to convert the charged liquid droplets into ultrafine/nanofibrous membranes. By adjusting the voltage the morphology and skeletal membrane parameters, such as porosity and shape/size distribution can be altered [4]. Fig. 10.1 gives different methods availbale in fabricating membranes.

10.2.3 Steps involved in desalination process

There are three stages in the desalination process, they are intake, reverse osmosis unit and posttreatment. Figs 10.2 and 10.3 illustarte different stages involved in desalination.

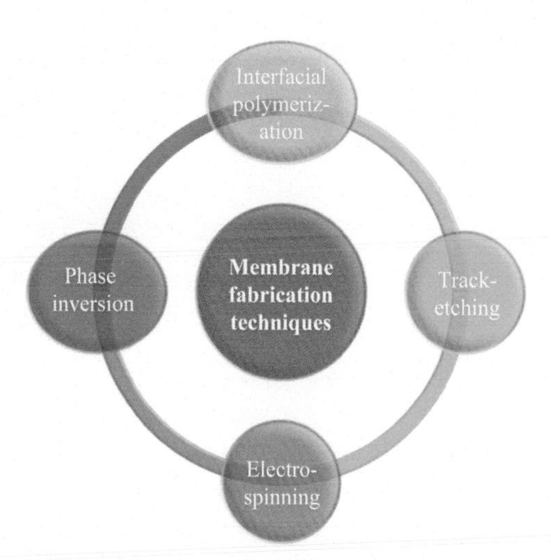

Figure 10.1 Different methods used in membrane fabrication.

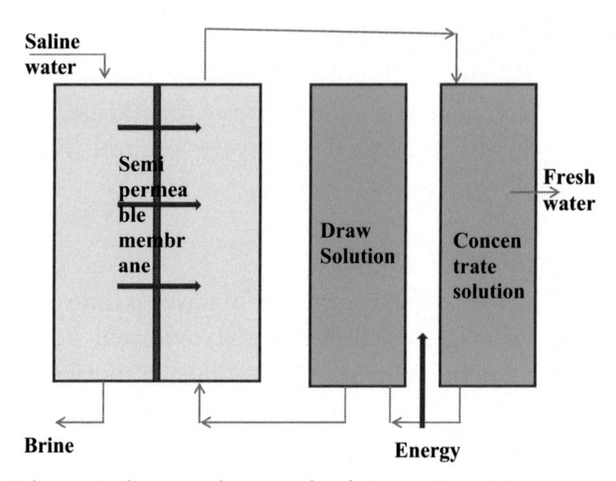

Figure 10.2 Schematic diagram showing desalination process.

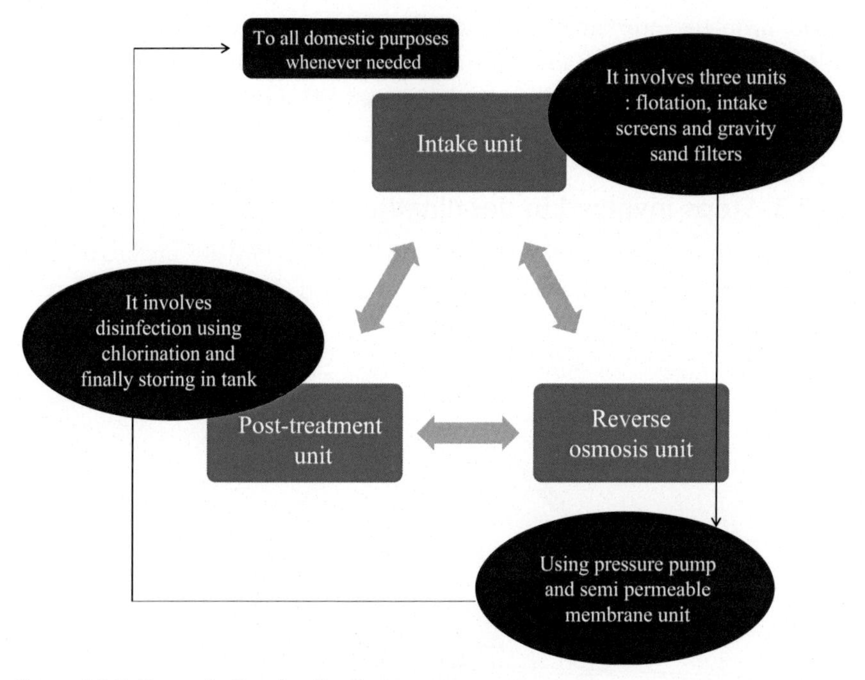

Figure 10.3 Stages in the desalination process.

10.2.4 Advanced materials involved in membrane removal

There are numerous materials available that are designed in order to fabricate nanoporous membranes that produce clear water by removing all pollutants from marine water. Properties like porosity, velocity distribution, permeation and water density are the parameters that are altered in such membranes compared to conventional membranes. This nanotechnology aids in the design of opportunistic energy efficient membranes, films and sheets for efficient pollutant removal. Some of the materials like graphene, zeolites and carbon nanotubes use mechanistic separation through reverse osmosis and capacitive desalination [4]. The materials that are used are listed below:

- Nanocarbon
- Nanometal oxides
- Graphene and its derivatives
- Zeolites
- Nanocomposites
- Carbon nanotubes

The inclusion of nanomaterials in a membrane improves the membrane permeability, fouling resistance, mechanical and thermal stability [3].

10.2.5 Different nanomembrane and its uses

10.2.5.1 Nanofibre membrane

This class of membrane removes microsized particles from water without any fouling. They are fabricated using the electrospinning process. These nanofibres contain a high surface area, porosity and form nanofibre mats with complex pore structures. These membranes can exclude bacteria and viruses from polluted water. They can be employed in pretreatment prior to ultrafiltration or reverse osmosis [3].

10.2.5.2 Nanocomposite membrane

These are promising filtration units which are fabricated from mixed matrices and surface functionalization. Mixed matrices are nanofillers like inorganic oxides and polymers which have a substantial surface area. Hydrophilic metal oxide nanoparticles (Al_2O_3, TiO_2 and zeolite), antimicrobial nanoparticles (nano-Ag and CNTs) and photocatalytic nanomaterials (bimetallic nanoparticles, TiO_2) are some of the nanomaterials used for such applications. The addition of mixed matrices reduces fouling. Further addition of metal oxide nanoparticles to polymeric membranes increases membrane surface hydrophilicity, water permeability or fouling resistance. Also they enhance the thermal and mechanical stability of polymeric membranes. The addition of mesoporous carbon in polymeric matrices improves the semipermeability nature, thereby applying it in reverse osmosis process [3].

10.2.5.3 Thin-film composite membrane

This is a new category of composite membrane prepared by interfacial polymerization. Nanomaterials, such as nanozeolates, nano-Ag, nano-TiO_2 and CNTs, were incorporated as nanoparticles into active thin layers of thin-film composite. They improve the interfacially polymerized layers. This technology is now commercially available from LG NanoH2O, Inc. They successfully conducted a field test of a thin-film composite element and reported that the water flux of the thin-film composite membrane was twice the flux of the polyamide membrane and achieved salt rejection of $>$99.7% [3].

10.2.5.4 Aquaporin-based or biologically inspired membranes

Aquaporins are pore-forming protein channels. They are ideal membranes for making efficient biomimetic membranes for water purification. They provide nanostructured selective barriers with good mechanical strength [3].

Though nanomaterials are developed to remove pollutants from water the challenge lies in commercialization. They are costly to fabricate in the laboratory itself. Additionally they are of nanoscale range risk assessment and management is a challenge. Research should understand the potential hazard of these materials in detail.

10.3 Powdered activated carbon pulsed blanket

Pulsazur is a lamellar clarifier launched by M/s. Degremont Technologies that uses adsorption on powder activated carbon in a pulsed sludge blanket reactor. It is used to eliminate dissolved organic matter and micropollutants from water. It is highly feasible and affordable because of its low operating cost [5].

This technology can be used by integrating it with conventional treatment or membrane treatment after settling and flotation before the sand filtration unit. Preconditioned water circulates through the system in a constant and uniform flow, from bottom to top, across the PAC bed. The efficiency of the process comes from the expansion followed by the settling of the PAC to maintain the homogenous expansion of the blanket. The continuous renewal of activated carbon guarantees long-lasting purification performance by suppressing any risk of saturation: the activated carbon bed, whose smoothing capacity is very important, adsorbs all the peaks of the essential of micropollutants [5].

Simplicity and savings [5]

- It operates without polymers.
- It consumes low energy around 8 Wh/m^3.
- There are no possible risks of abrasion or corrosion.
- Highly economic because of the use of activated carbon.
- Optimal use of powdered activated carbon is kept homogenous by upflow pulsation.
- It is highly flexible as it supports high variation in loads and quality of water.
- Hydraulics are designed in such a way that they maintain velocity within the system.
- Complete removal of micropollutants and dissolved organic matter [5]

10.4 Role of bioemulsificant on oil spill removal

Microbes produce their own surfactants, known as biological compounds, which consist of sugars, amino acids and lipids. These compounds are classified into two groups: biosurfactants which have low molecular weight compounds (Rhamnolipids and surfactin produced by *Pseudomonas aeruginosa* and *Bacillus subtilis*); and bioemulsifiers comprising high-molecular-weight compounds (emulsan and alasan produced by *Acinetobacter radioresistens* and *Acinetobacter venetianus* recombination activating genes). An alternative green way to clean marine oil spill is bioremediation through such biological compounds [6]. One of the main reasons for the availability of hydrocarbons in the environment is their low solubility in water, therefore they are absorbed on the soil surface, thereby limiting its availability to biodegrading microorganisms. The removal of oil spill through bioremediation is the interaction between the host microorganism and the hydrocarbon chain. The reaction may not be complete due to some interruptions in environmental factors and the low solubility of oil in water. Hence apparent solubility is increased through the addition of emulsifier molecules. Emulsifier molecules may be of chemical or biological origin [7]. A bioemulsifier is a high-molecular-weight compound that enhances oil recovery and hydrocarbon bioremediation in an aquatic environment. These molecules reduce the surface and interfacial tensions in both aqueous solutions and hydrocarbon mixture and have very low (l g/mL) critical micelle concentrations (CMC) that increase the apparent solubility and increase the availability of the oil to the microbial attack [7]. They possess certain advantages such as low toxicity, biodegradability and effectiveness in a wide range of pH and temperature compared to other chemical/synthetic surfactants. Synthetic surfactants are costlier than these natural compounds [7].

Synthetic surfactants are mainly used as dispersant agents that break oil slicks into smaller droplets and can move easily in the water column. Synthetic dispersants Corexit 9527 and Corexit 9500 A comprising anionic surfactant dioctyl sodium sulfosuccinate (DOSS) were widely used for the Deepwater Horizon oil spill response. They are chemically stable but not biodegradable as they persist longer in the marine environment [6]. The authors used bioemulsifier exopolysaccharide EPS_{2003} along with microbial communities for degrading oil spills in water. The study kinetics were assessed using a microcosm study. The data obtained from the microcosm experiment indicated that EPS_{2003} could be used for the dispersion of oil slicks and could stimulate the selection of marine hydrocarbon degrading agents, thus increasing the bioremediation process [7].

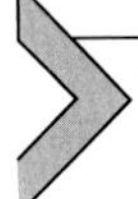

10.5 Oil spill removal using skimmers by National Ocean and Atmospheric Administration

10.5.1 Skimming

Skimming is the process of removal of oil from water before it reaches the sensitive areas of the coastline. The success of this process relies on the encounter rate which depends on direct contact of the skimmer with oil in order to remove it from the seawater's surface. Sometimes two boats will tow a collection boom allowing concentration within the boom. Ideal conditions for skimming are during the day when the oil slick and ocean surface is fairly calm. Skimming devices attract oil to their surface before transfer to a collecting tank [8].

10.5.2 In situ burning

This is the process of burning the spilled oil on the ocean. Generally burning is used for treating oiled marshes. The success of this process lies on corralling a layer of oil thick enough to maintain a sustained burn. Similar to skimming, two boats will often tow a fire-retardant collection boom to concentrate enough oil to burn [8].

10.5.3 Chemical dispersant

This is the process of breaking oil into small droplets using certain chemicals. These droplets will be readily available for microbes to degrade them completely. But they pose serious impacts on marine life. Ideal conditions for chemical dispersion are daylight with mild winds and moderate seas [8].

10.6 Marine acts and regulations

This section covers some of the Acts that the Ministry of Environment is responsible for in preventing marine pollution.

10.6.1 Exclusive Economic Zone and Continental Shelf (Environmental Effects) Act 2012

The Act aims to promote the sustainable management of natural resources in the exclusive economic zone (EEZ) and continental shelf. Also it seeks to protect the EEZ and continental shelf from pollution by regulating discharges and dumping. Further it seeks to achieve these goals by allowing for the regulation of certain activities that were previously unregulated in the EEZ and continental shelf [9].

10.6.2 Regulations under the Exclusive Economic Zone and Continental Shelf (Environmental Effects) Act 2012

They cover regulations classifying activities that are as permitted. They are:

- Seismic surveying
- Submarine cabling
- Marine scientific research
- Prospecting and exploration phases of seabed mineral mining and petroleum

It also focuses on regulations for cost recovery [9].

10.6.3 Resource Management (Marine Pollution) Regulations 1998

These Regulations control dumping and discharges from ships and offshore installations in the coastal marine area. The Regulations deal with the dumping of waste and discharges from vessels including oil, garbage and sewage [9].

10.7 Marine pollution removal using new innovations as reported ideas

10.7.1 Cleaning up oil spills with magnets and nanotechnology

Oil spills from container ships or offshore platform are very frequent hazards to marine and coastal ecosystems and are very expensive to clean. In order to address this issue researchers from the Massachusetts Institute of Technology (MIT) have found a method of recovering oil after a spill using magnets, potentially saving companies like BP money in clean-up bills in 2012. As we all know oil is not magnet, but when mixed with water-repellent nanoparticles that contain iron, the oil can be magnetically separated from the water. The nanoparticles can later be removed to enable the reuse of the oil. The process of removal is that seawater which is polluted due to oil will be pumped onto a boat treatment facility. Once onboard the magnetic nanoparticles will be added and attach themselves to oil thereby removing them [10].

The liquid would then be filtered with the magnets to separate the oil and water, with the water returned to the sea and the oil carried back to shore to an oil refinery. This solution has been developed since the 2010 Gulf of Mexico oil spill and there has been a rise in interest from oil companies and government departments for funding new techniques for reducing the environmental impact and cost of future oil spills. Many scientists project that skimmers are the best option when it comes to the large scale instead of magnets, which are best suited for small-scale spills. Accordingly a skimmer designed by Illinois company Team Elastec offered better results than anything. It can recover about 4700 gallons per minute, so assuming the skimmer could be deployed 24 h a day, it would take 30 days to pick up the entire 200 million gallons of oil spilled during the Gulf of Mexico disaster [10].

10.7.2 Illinois team Oil spill cleanup using skimming option

Team Elastec, an Illinois-based company that is a veteran in the oil spill clean-up business, developed giant grooved discs that skimmed oil more than three times better than the industry standard. It showed an efficiency of about 89.5% with high removal rate around 17,677 L/min [11].

10.8 Conclusion

This chapter covers some of the novel technologies that are used in removing marine pollutants to the fullest. All the technologies discussed here are highly recommended because they show unique properties. Technologies like the use of nanomaterials paved way to the direct smart use of advanced materials to remove heavy metals, micropollutants and other debris like microorganisms. Though such innovations offer high efficiency their success lies in commercialization that depends on cost–benefit analysis. Additionally this chapter discussed some new techniques like chemical dispersant, skimmers and pulsed blanket reactor with powdered activated carbon in removing organic matter and heavy metals. But all these strategies rely on the valuable rational-designing idea for assorted materials owing to environmental utilities.

References

[1] Humplik T, Lee J, O'Hern SC, Fellman BA, Baig MA, Hassan SF, Atieh MA, Rahman F, Laoui T, Karnik R, Wang EN. Nanostructured materials for water desalination. Nanotechnology 2011;22(29):292001. Available from: https://doi.org/10.1088/0957-4484/22/29/292001.

[2] Jordan E, Jessica Martin, Hakeem MH, Suryajaya T, Nugraha T, Listyorin NT. A review of nanotechnology application for seawater desalination process. Surya Octagon Interdiscip J Technol 2016;1(2):155–79.

[3] Kunduru KR, Nazarkovsky M, Farah S, Pawar RP, Basu A, Domb AJ. Nanotechnology for water purification: applications of nanotechnology methods in wastewater treatment. Water Purif 2017;33–74. Available from: https://doi.org/10.1016/b978-0-12-804300-4.00002-2.

[4] Dongre RS. Rationally fabricated nanomaterials for desalination and water purification. In: Novel Nanomaterials - synthesis and applications. <https://doi.org/10.5772/intechopen.74738>; 2018.

[5] Suezwater handbook, Degremont R Technologies, drinking water production, treatment of activated carbon, powder activated carbon contactor Pulsazur. <https://www.suezwaterhandbook.com/degremont-R-technologies/drinking-water-production/treatment-on-activated-carbon/powder-activated-carbon-contactor-Pulsazur>.

[6] Doshi B, Sillanpaa M, Kalliola S. A review of bio-based materials for oil spill treatment. Water Res 2018;135:262–77.
[7] Cappello S, Genovese M, Torre CD, Crisari A, Hassanshahian M, Santisi S, Calogero R, Yakimov MM. Effect of bioemulsificant exopolysaccharide (EPS_{2003}) on microbial community dynamics during assays of oil spill bioremediation: a microcosm study. Mar Pollut Bull 2012;64:2820–8.
[8] NOAA, response and restoration, media, how to remove oil spill from seawater. <www.noaa.gov>.
[9] Marine, Marine acts and regulations. <www.mfe.govt.nz>.
[10] <https://edition.cnn.com/2012/09/21/tech/oil-spill-magnets/index.html>.
[11] <www.nationalgeographic.com>.

Case study on marine pollutants and its impacts

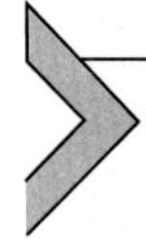

Case study 1: Evaluating the ocean cleanup in cleaning marine debris in North Pacific Gyre using SWOT analysis

The work has been carried out by Morrison, E., Shipman, A., Shrestha, S., Squier, E., & Stack Whitney, K. (2019).

Reference: "Morrison E, Shipman A, Shrestha S, Squier E, Stack Whitney K. Evaluating The Ocean Cleanup, a Marine Debris Removal Project in the North Pacific Gyre, Using SWOT Analysis. Case Studies in the Environment; 2019. doi:10.1525/cse.2018.001875"

Introduction

Marine debris comes from various sources caries wood, plastics, metals, and Styrofoam. The major effect by these pollutants is that they have the capacity to leach into seawater. Pollutants that are commonly found in the "Great Garbage Patch" in the North Pacific gyre include carcinogens, pesticides, heavy metals, and endocrine disruptors. The concentration of these pollutants exceeds 1000 parts per billion which is extremely high for seawater. This case study address how plastics ends up in ocean as debris, impacts on human health, wildlife conservation, and economic aspects in the location chose.

SWOT analysis

It is a strategic planning technology for projects in the organisation and institution. SWOT stands for Strength, Weakness, Opportunities, Threats. This analysis provides the comprehensive advantages and disadvantages on the study addressed. It is applied in many fields including environmental planning and conservation.

Case examination

Cleaning up of ocean that has plastic pollution is a critical issue because it involves multiple time- and spatial scales. Like beach cleanup it has taken place for a decade and is organised at local and regional bodies. Ocean cleanup was founded by Boyan Slat in 2013 with the goal of designing and cleaning marine debris from the ocean. There are several advantages and disadvantages to this ocean cleanup which are studied using this technology to achieve its success.

Strengths

There are two strengths in using this idea in cleaning the North Pacific Gyre that is polluted by marine debris.

- The ocean cleanup is the first organization attempting to clean marine debris at a large scale. The North Pacific Gyre has the greatest accumulation of floating trash in the world, around 1.8 trillion plastics according to the survey report. The design is a U-shaped system that can float in the gyre to collect floating plastics. The collection system will use the force of the oceanic gyre to direct debris into a barrier.
- The organization has learnt from many previous experiences with failed designed prototypes, hence the proposed design will achieve its goal in a successful way.

Weakness

- The first full-scale deployment was not successful. The main agenda was to propose a 600-m-long system and to engage it in cleaning that works with supporting vessels to extract the debris for every 6 weeks. But at the end of December 2018 the debris was not successfully collected but rather it was brought back to shore after breakage.
- It is unclear how it will impact the marine life. There are possibilities for the system to remove many floating organisms like neuston that live on the surface of the ocean water.
- This system does not reduce the overall production or consumption of plastics.

Opportunities

- There is a global need to address the cleaning of marine debris and its harms to marine life. Major harm is eating of plastics by wildlife in marine biota. They attack the food chain through indirect ways.
- Cleaning up marine debris may help in improving the economy and prevents impacts reaching human health.
- Citizen science and community-led initiatives do not appear to be effective in cleaning marine debris. The Georgia Sea turtle centre marine debris initiative (GSTCMDI),which involved citizens in cleaning marine debris, together collected 6527 pieces in 461 hours.

Threats

- There are chances for other organizations to build better systems than cleanup. Like Korea fisheries infrastructure promotion association operates a vessel that can remove 350 tons of marine debris annually.
- Another important monitoring technology is the use of satellite location tracking buoys that are used to monitor the movement of debris. This helps the researchers to identify where the debris is located and has traveled.

Conclusion

This analysis helped to find the merits and demerits of any technologies that are used by or initiated by any organization. This can also be extended to compare alternative approaches such as coastal containment of plastics

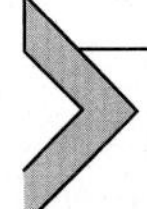

Case study 2: Microplastics pollution after the removal of the Costa Concordia wreck: first evidence from a biomonitoring case study

Reference: "Avio CG, Cardelli LR, Gorbi S, Pellegrini D, Regoli F. Microplastics pollution after the removal of the Costa Concordia wreck: First evidences from a biomonitoring case study. Environ Pollut 2017; 227: 207–214. doi:10.1016/j.envpol.2017.04.066"

The case study deals with the accumulation of microplastics in marine life at different zones. Ingestion of microplastics depends on parameters, for example, size, shape, and density, that determine the position of particles in the water column or sediments, and hence their availability for marine organisms. It is found that lots of microplastics had been detected in wild caught and commercial fish species from many parts of the coast. These microplastics are polystyrene, polyamide, nylon, polypropylene, polyester, and polyurethane. Microplastics are defined as emerging pollutants and the European Marine Strategy Framework Directive (MSFD, 2008/56/EC) included marine litter and microplastics among the descriptors of the Good Environmental Status.

This case study purely denotes the availability of microplastics in different benthic fishes sampled after 2.5 years of huge engineering operation for the parbuckling project on the Costa Concordia wreck at Giglio Island. This operation required the use of 30,000 tons of steel (equivalent to four Eiffel Towers), 21 pylons of more than 1 m diameter drilled for 9 m in the granite sea-bottom of the island to fix six artificial platforms, 56 chains for the anchoring system each 58 m long and 26 tons in weight, and 1189 cement grout bags with a total volume of 12,000 m^3 and more than 16.000 tons weight. Nearly 30 vessels and aircraft were used with 500 workers onsite. This denotes anthropogenic pressure on this area for completing this project in site might have released pollutants in the area. The National Civil Protection and the Italian Institute for Environmental Protection and Research (ISPRA) coordinated a large monitoring program, excluding serious contamination events from the wreck or a consistent increase of chemical pollution in this area. It was the self-interest of the authors to identify the pollutants in the location.

A total of 41 fish representative of different commercial species were sampled from two areas of Giglio island in September 2014 during the final operations for the removal and towing away of the Costa Concordia wreck. Some of the common species collected in the two areas included *Scorpaena* sp., *Uranoscopus scaber*, and *Phycis phycis* as typically benthonic fishes, and *Spondyliosoma cantharus* as a benthopelagic species. Due to the lack of a wild mussel population on the Island, an active mussel watch approach was performed caging these organisms at different depths of the water column in the wreck and a control site. Two caging sites were selected, one north of the Costa Concordia wreck and another in front of the Caldane beach. Two different depths were used: 1.5 m from surface and 30–35 m from the bottom, for incubation and finally samples were dissected. The polymers were extracted from gastrointestinal tracts of fish and soft tissue of mussels using trituration of dried samples followed by separation under density gradient and filtration under vacuum, partial digestion in 15% H_2O_2, visual sorting, and FT-IR characterization.

The results showed that the extraction of gastrointestinal tracts showed the presence of ingested microplastics from ocean water. On average, the typically benthonic species, *Phycis phycis, Scorpaena* sp., and *Uranoscopus scaber*, showed plastic particles in 77%, 84%, 86% of microplastics, respectively, while the benthopelagic *Spondyliosoma cantharus* exhibited microplastics in 100%. The size of plastics items ingested by fish exhibit some common size classes ranging between 0.5 and 1 mm (37%), 0.1–0.5 mm (35%), followed by 1–5 mm (20%), and smaller than 0.1 mm (8%). FTIR analysis confirmed that polyethylene was the most common polymer found in the sampled fishes. Shapes of plastics are fragments, lines, and films in the sampled species. The results from caged mussels denote the presence of plastics varied from 1–2 items/individuals depending on the depth. Surface-caged mussels contained particles in the size classes 0.1–0.5 mm, 0.5–1 mm, and 1–5 mm, whereas mussels caged in proximity to the bottom were largely dominated by the 1–5 mm size class.

Thus this study reveals a consistent increase of microplastic pollution in the benthic environment caused by the huge operations for the removal of the Costa Concordia wreck. The study also denotes the high frequency of microplastics in fishes and limited capacity of mussels for microplastics. This clearly indicates the impact on the benthic environment and on the seawater column at the end of removal activities.

APPENDIX 1

List of abbreviations

AVHRR advanced very high-resolution radiometer
BWM ballast water management
CDOM colored dissolved organic matter
CNT carbon nanotubes
DDT dichlorodiphenyltrichloroethane
DNA deoxyribonucleic acid
EDS energy dispersive X-ray spectroscopy
EPA environmental protection agency
ESA endangered species act
GO graphene oxide
HAB harmful algal blooms
HMW high molecular weight
IMO international maritime organisation
IPM integrated pest management
MEPC Marine Environment Protection Committee
MODIS moderate resolution imaging spectroradiometer
NAP nonalgal pigments
NCM nanocomposite membranes
NF nanofiltration
NOAA National Ocean and Atmospheric Administrative
PAC powdered activated carbon
PAH poly aromatic hydrocarbons
PCB polychlorinated bisphenyl
PVDF polyvinylidene difluoride
RNA ribonucleic acid
RO reverse osmosis
SEM scanning electron microscopy
SLSTR sea and land surface temperature radiometer
SS suspended solids
SST sea surface temperature
SWRO sea water reverse osmosis
TFC thin film composite
USEPA United States of Environmental Protection Agency
UV ultraviolet radiation
VIIRS visible infrared imaging radiometer suite

APPENDIX 2

List of symbols

Cl^{-} Chloride ion
Na^{+} Sodium ion
SO^{2-} Sulfate ion
Mg^{2+} Magnesium ion
Ca^{2+} Calcium ion
K^{+} Potassium ion
CO_2 Carbon dioxide
^{3}H Tritium
^{14}C Radio carbon
^{90}Sr Strontium
^{99}Tc Technetium
^{137}Cs Cesium
^{210}Po Polonium
^{241}Pu Plutonium
^{241}Am Americium
Al_2O_3 Aluminium oxide
TiO_2 Titanium dioxide

Glossary

Adsorption process by which solid substance holds molecules of gas or liquid as thin film

Agglomeration An assemblage of molecules or collection of molecules

Advection it is the process of transfer of heat or matter by flow of fluid horizontally in the sea or in atmosphere

Aerobic reaction in the presence of oxygen

Anaerobic reaction in the absence of oxygen

Bioaccumulation it is the accumulation of pollutants gradually in the environment

Biodegradation it is the process of natural breakdown of materials by microorganisms

Bioluminescent it is production and emission of light by living microorganisms

Bioremediation it is process of use of naturally occurring or deliberately introduced microorganism for the breakdown of environmental pollutants

Biosensors it is a device used to detect pollutants using microorganism as major detector

Biotransformation it is the process of alteration of chemicals in the environment under the influence of external factors

Coagulation it is the process by which small suspended particles forms flocs using external chemicals like coagulants

Dispersion it is the process of distributing substances or chemicals evenly in the medium

Electrocoagulation it is the process of removal of pollutants under the influence of electric charge with respect to coagulants thereby changing the surface particle charge allowing suspended matters to form agglomerates

Eutrophication it is the process of excessive growth of algae due to enriched availability of nutrients thereby depleting dissolved oxygen in water

Filtration it is physical or chemical or biological operation that separates solid matter from water through filtering medium

Flocculation it is the process of formation of flocs in the water carrying suspended matters

Hydrolysis it is a chemical reaction involving breaking of water molecule

Immobilization it is the process of immobilizing microbes or enzymes in an inorganic substances in order to retain its activity

Incineration it is the process of conversion of organic substance to gaseous nature in an closed controlled environment under the influence of heat

Microbial degradation it is the process of mineralization of any xenobiotic compound through microbes

Mycodegradation it is the process of mineralization of any xenobiotic compound through fungi in the environment through serious of chemical reaction

Ocean acidification alteration in the pH of sea water through changes in organic or inorganic chemicals due to disturbance created by external matters

Ozonation degradation of chemical compounds using ozone as an active substance

Photodegradation degradation of any complex chemical compounds using sunlight or artificial light source

Phytoremediation it is the process of removal of polluting chemicals using plants

Reverse osmosis water purification process that uses a partially permeable membrane to remove ions, unwanted molecules and larger particles from drinking water

Sonication it is process of use of ultrasonic sound waves for complete removal of pollutants from water

Thermal degradation it is involvement of excessive heat for degrading any chemical compounds from water

Vermiculation use of worms like earthworms for breakdown of pollutants in any solid medium.

Index

Note: Page numbers followed by "*f*" and "*t*" refer to figures and tables, respectively.

Printed in the United States
By Bookmasters